# SpringerBriefs in Systems Biology

For further volumes:
http://www.springer.com/series/10426

Emily G. Armitage • Helen L. Kotze
Kaye J. Williams

# Correlation-based network analysis of cancer metabolism

A new systems biology approach in metabolomics

 Springer

Emily G. Armitage
Universidad San Pablo CEU, Centre for
Metabolomics and Bioanalysis
Boadilla del Monte
Madrid
Spain

Kaye J. Williams
University of Manchester, Manchester
Pharmacy School, Manchester Cancer
Research Centre
Manchester
United Kingdom

Helen L. Kotze
University of Manchester Manchester
Institute of Biotechnology
Manchester
United Kingdom

ISSN 2193-4746            ISSN 2193-4754 (electronic)
ISBN 978-1-4939-0614-7    ISBN 978-1-4939-0615-4 (eBook)
DOI 10.1007/978-1-4939-0615-4
Springer New York Heidelberg Dordrecht London

Library of Congress Control Number: 2014933544

Printed on acid-free paper

Springer is part of Springer Science+Business Media (www.springer.com)

# Contents

# Chapter 1
# An Overview of Cancer Metabolism

The metabolome is considered the closest entity to the phenotype of a biological system. It displays the changes made at higher hierarchical levels such as the proteome, transcriptome and genome. For many diseases including cancer, studying the metabolome enables us to gain a better understanding of global biological response of cancer cells in the progression of the disease. Revealing the complexity of the metabolome is particularly advantageous to understand the phenotypic function of a cancer cell that is governed by the preceding levels (proteins, transcription factors and genes).

Cancer metabolism has been studied for decades, revealing cancer cell function in order to provide an insight into the disease. Differences in central carbon metabolism between cancerous and normal cells were first demonstrated by Otto Warburg in the 1920s (Warburg et al. 1927). He evaluated the metabolic consumption of glucose and found that cancer cells preferentially used glycolysis over oxidative phosphorylation even in the presence of oxygen (Warburg 1956). This so-called "Warburg effect" is frequently observed in many cancer types, although the underlying basis and consequence of this phenomenon are still not wholly clarified and there appears to be no single mechanism that drives an aerobic tumour cell towards a glycolytic phenotype. Indeed, it is likely that there is plasticity in how a cancer cell metabolises glucose, dependent on glucose availability and the local cellular microenvironment among other potential influencing factors. Furthermore, elevated glucose levels can suppress both glycolysis and oxidative phosphorylation via the "Crabtree effect", which is generally accepted as a short-term, reversible response to glucose availability (Diaz-Ruiz et al. 2011).

From the perspective of ATP generation alone, a reliance on glycolysis vs. oxidative phsophorylation makes little sense. However, the main hypothesis ventured for why the Warburg effect benefits cancer cells is focused less on ATP generation per se and more on coincidental generation of the cellular building blocks required in rapidly proliferating cells. Glycolysis can provide intermediary precursors that feed into many biosynthetic pathways that ultimately generate nucleotides, amino acids and lipids as well as ATP.

This is achieved by multiple mechanisms. Cancer cells generally upregulate the expression of glucose transporters (predominantly Glut-1), enabling enhanced

E. G. Armitage et al., *Correlation-based network analysis of cancer metabolism*,
SpringerBriefs in Systems Biology, DOI 10.1007/978-1-4939-0615-4_1,

glucose uptake into cells, a phenomenon that is exploited in positron emission tomography using deoxyglucose labelled with the positron-emitting $^{18}$F isotope ($^{18}$F-FDG PET). Downstream of the glucose transporters, glycolytic enzyme expression is modified to increase flux, achieved either by elevated protein expression and/or expression of isoforms pertaining to altered activity within the cancer cell environment. Importantly, documented changes include those in enzymes recognised as the rate-controlling steps in glycolysis: hexokinase (HK), phosphofructokinase (PFK) and pyruvate kinase (PK), which have been previously reviewed (Diaz-Ruiz et al. 2011). In the case of PK, cancer cells often express the M2 isoform. Within the glycolytic pathway, PK catalyses the conversion of phosphoenolpyruvate to pyruvate, a process that generates ATP. The M2 variant is actually less effective at catalysing this reaction, which although may compromise ATP production (Mazurek et al. 2005), is beneficial to the potential shunting of upstream metabolites into biosynthetic pathways (Christofk et al. 2008a, b).

Multiple factors contribute to the observed metabolic changes in cancer cells. From a genetic perspective, activation of key oncogenes and loss of tumour suppressor function can drive perturbed metabolism. Oncogenic transformation of c-myc or mutations leading to activation of signalling via the phosphoinositide 3-kinase (PI3K)/Akt (Protein kinase B, PKB)/the mammalian target of rapamycin (mTOR) pathway are commonly associated with increased glucose metabolism in cancer cells. Activation of c-myc leads to the enhanced expression of multiple genes involved in glycolysis, in particular, lactate dehydrogenase (LDH) (Shim et al. 1997), whereas the PI3K/Akt/mTOR axis is commonly associated with direct enhancement in glucose uptake via regulation of transporter activity, although there will undoubtedly be crossover in function given the ability of both pathways to drive expression of the transcription factor hypoxia inducible factor-1 (HIF-1, refer to Chap. 2 for further information; Bardos and Athcroft 2004; Dang et al. 2008; Burrows et al. 2010). mTOR is negatively regulated by AMP-activated protein kinase (AMPK). AMPK is activated by the antidiabetic drug metformin (Zhou et al. 2001), leading to intriguing possibilities for targeting cancer metabolism as a preventative or therapeutic approach with established drugs (Pierotti et al. 2013).

Tumour suppressors linking to glycolytic phenotype include phosphatase and tensin homolog (PTEN), p53, tuberous sclerosis 1 and 2 (TSC1 and 2) and liver kinase B1 (LKB-1). PTEN is a classical inhibitor of the PI3K pathway, however, genetic over expression of PTEN, suppresses the Warburg effects via both PI3K-dependent and -independent effects, the latter including inhibition of c-myc (Garcia-Cao et al. 2012). p53 elicits multiple effects on metabolism (Vousden and Ryan 2009). Glucose uptake is suppressed via inhibiting the expression of glucose transporters. Glycolytic flux is downregulated through reduced expression of the glycolytic enzyme phosphoglycerate mutase and upregulation of the protein TP53-induced glycolysis and apoptosis regulator (TIGAR), which lowers fructose-2,6-bisphosphate levels in cells through the inhibition of PFK activity (Bensaad et al. 2006). TSC1 and 2 exert their effects via suppression of mTOR activity (Lee et al. 2007) and LKB1 via activation of AMPK (Shaw et al. 2004).

Although enhanced flux through the earlier stages of the glycolytic pathway can be seen as beneficial to generate precursors, exactly why, in the presence of oxygen, pyruvate would be converted to lactate as opposed to entering oxidative phosphorylation is a matter of continued debate. One clear advantage is that the pyruvate-to-lactate conversion that is catalysed by LDH regenerates NAD+, which is required for earlier stages in glycolysis. Indeed, LDH is commonly overexpressed in tumours and the high enzyme activity may reduce the amount of pyruvate available to enter mitochondria. Further contributors to the reduced use of pyruvate in mitochondrial metabolism may be impaired transport of the metabolite through the mitochondrial membrane and reduced activity of pyruvate dehydrogenase (PDH) complex, which catalyses the irreversible oxidative decarboxylation of pyruvate to acetyl CoA that feeds into the Krebs cycle. Both have been reported in cancer cells (Diaz-Ruiz et al. 2011). In addition to impairments in how pyruvate is utilised, defects in the Krebs cycle have also been suggested. Several intermediates of the Krebs cycle are also precursors for biosynthetic pathways. In rapidly dividing tumour cells, the increased diversion of Krebs intermediaries towards biosynthesis (for example citrate towards lipid biosynthesis) could compromise oxidative metabolism. Furthermore, tumour cells have been shown to exhibit mutations in several enzymes of the Krebs cycle, including isocitrate dehydrogenase 1(IDH1), succinate dehydrogenase (SDH) and fumarate hydratase (FH) that compromise or modify enzyme function. Both SDH and FH are recognised tumour suppressors. Germline mutations of SDH expression predispose to phaeochromocytomas and paragangliomas, whereas FH deficiency predisposes to leiomyomatosis and renal cell cancer. Acquired mutations in the genes encoding all three enzymes are observed in multiple cancer types (Bardella et al. 2011; Chen and Russo 2012). FH deficiency in renal cell carcinomas drives transition to a glycolytic phenotype associated with a reduction in AMPK levels (Tong et al. 2011). In addition to cancer cells possessing defects in the Krebs cycle, oxidative phosphorylation can also be directly suppressed via reduced ADP uptake into the mitochondrial matrix (Chan and Barbour 1983) and reduced activity of ATP synthase (Cuezva et al. 2007).

Inspite of the Warburg effect and the above mentioned defects within the Krebs cycle and mitochondrial metabolism, cancer cells do often maintain significant levels of oxidative phosphorylation. One potential contributing factor is the use of substrates other than glucose to fuel metabolism. High uptake of the amino acid glutamine has been observed in many cancer cell lines, which is coupled to elevated glutaminolysis (Matsuno and Hirai 1989; Matsuno and Goto 1992) via the sequential activities of glutaminase and glutamate dehydrogenase or transaminases, such that $\alpha$-ketoglutarate is generated. This can then feed into the Krebs cycle in an overall process known as glutamine anaplerosis (DeBerardinis et al. 2008). Recent evidence has shown that the activity of glutamate dehydrogenase is regulated by components of the mTOR pathway (Csibi et al. 2013), linking this axis to elevated glucose and glutamine consumption in cancer cells. In addition to fuelling the Krebs cycle, glutaminolysis also supports a number of biosynthetic pathways in cancer cells. Interestingly, a further product of the reactions is lactate, thereby raising the possibility that the high levels of lactate in cancer cells arise not only from anaerobic glycolysis manifested via the Warburg effect but also via the metabolism of glutamine (DeBerardinis et al. 2007).

Lactate metabolism by cancer cells has received a lot of interest in recent years. Although predominantly thought of as a waste product of anaerobic glycolysis, recent evidence suggests that cancer cells can actually utilise lactate to fuel oxidative phosphorylation in the presence of oxygen. Indeed, an effective "metabolic symbiosis" has been observed between "lactate-producing" and "lactate-using" cells within tumours (Sonveaux et al. 2008). Cells can be driven towards a "producer" or "user" phenotype dependent upon the microenvironmental conditions in which they prevail within the tumours. Indeed, a "reverse Warburg effect" has been described (Bonuccelli et al. 2010) in which stromal cells within the tumour (cancer associated fibroblasts) are actually the glycolytic, "lactate-producing" moiety, feeding the "lactate-using" cancer cells. Lactate transport is governed by the monocarboxylate transporters MCT-1 and MCT-4, with MCT-4 predominantly functioning as an "exporter" and MCT-1 as an "importer" in the proposed models of cooperativity (Sonveaux et al. 2008). Experiments using labelled metabolites show [$^{13}$C]lactate conversion to [$^{13}$C]glutamate via the Krebs cycle. Lactate is initially converted to pyruvate via LDH. There are then two routes for entry into the Krebs cycle. One via PDH to generate acetate and subsequently glutamate following a spin-off from α-ketoglutarate, and one via anaplerosis whereby pyruvate carboxylase converts pyruvate into oxaloacetate. Lactate metabolism is not restricted to cancer cells, being observed also in the brain, for example (Gallagher et al. 2009). Furthermore, similarities in the reverse Warburg effect have been observed along with metabolic cooperativity between neurons and glia cells within the brain (Pavlides et al. 2010).

These exciting new observations in cancer cell metabolism have reinvigorated research into metabolism as a therapeutic target for cancer. As discussed above, metabolic alterations in cancer are observed across glycolysis, the Krebs cycle and oxidative phosphorylation. There is heavy reliance on using alternative substrates and on anaplerotic reactions to replenish Krebs cycle intermediates that can be exploited in cancer therapy. The oncogenic drivers of metabolism along the PI3K have been targeted by small molecule approaches, and many are in clinical evaluation. Direct targets within cancer metabolism based on our current knowledge have recently been reviewed and include glucose transporters, HK and MCTs (Galluzzi et al. 2013). However, with technological advances now enabling the application of metabolomic and network-based correlation approaches to cancer metabolism, these targets are likely to be only the tip of the iceberg.

# References

Bardella C, Pollard PJ, Tomlinson I (2011) SDH mutations in cancer. Biochim Biophys Acta 1807(11):1432–1443. doi:10.1016/j.bbabio.2011.07.003
Bardos JI, Athcroft M (2004) Hypoxia-inducible factor-1 and oncogenic signalling. Bioessays 26(3):262–269. doi:10.1002/bies.20002

Bensaad K, Tsuruta A, Selak MA, Vidal MNC, Nakano K, Bartrons R, Gottlieb E, Vousden KH (2006) TIGAR, a p53-inducible regulator of glycolysis and apoptosis. Cell 126(1):107–120. doi:10.1016/j.cell.2006.05.036

Bonuccelli G, Whitaker-Menezes D, Castello-Cros R, Pavlides S, Pestell RG, Fatatis A, Witkiewicz AK, Vander Heiden MG, Migneco G, Chiavarina B, Frank PG, Capozza F, Flomenberg N, Martinez-Outschoorn UE, Sotgia F, Lisanti MP (2010) The reverse Warburg effect glycolysis inhibitors prevent the tumor promoting effects of caveolin-1 deficient cancer associated fibroblasts. Cell Cycle 9(10):1960–1971

Burrows N, Resch J, Cowen RL, von Wasielewski R, Hoang-Vu C, West CM, Williams KJ, Brabant G (2010) Expression of hypoxia-inducible factor 1 alpha in thyroid carcinomas. Endocr Relat Cancer 17(1):61–72. doi:10.1677/erc-08-0251

Chan SHP, Barbour RL (1983) Adenine-nucleotide transport in hepatoma mitochondria—charaterization of factors influencing the kinetics of ADP and ATP uptake. Biochimica Biophysica Acta 723 (1):104–113. doi:10.1016/0005-2728(83)90014-2

Chen JQ, Russo J (2012) Dysregulation of glucose transport, glycolysis, TCA cycle and glutaminolysis by oncogenes and tumor suppressors in cancer cells. Biochim Biophy Acta 1826(2):370–384. doi:10.1016/j.bbcan.2012.06.004

Christofk HR, Vander Heiden MG, Harris MH, Ramanathan A, Gerszten RE, Wei R, Fleming MD, Schreiber SL, Cantley LC (2008a) The M2 splice isoform of pyruvate kinase is important for cancer metabolism and tumour growth. Nature 452(7184):230–274. doi:10.1038/nature06734

Christofk HR, Vander Heiden MG, Wu N, Asara JM, Cantley LC (2008b) Pyruvate kinase M2 is a phosphotyrosine-binding protein. Nature 452(7184):181–127. doi:10.1038/nature06667

Csibi A, Fendt SM, Li CG, Poulogiannis G, Choo AY, Chapski DJ, Jeong SM, Dempsey JM, Parkhitko A, Morrison T, Henske EP, Haigis MC, Cantley LC, Stephanopoulos G, Yu J, Blenis J (2013) The mTORC1 pathway stimulates glutamine metabolism and cell proliferation by repressing SIRT4. Cell 153(4):840–854. doi:10.1016/j.cell.2013.04.023

Cuezva JM, Sanchez-Arago M, Sala S, Blanco-Rivero A, Ortega AD (2007) A message emerging from development: the repression of mitochondrial beta-F1-ATPase expression in cancer. J Bioenerg Biomembr 39(3):259–265. doi:10.1007/s10863-007-9087-9

Dang CV, Kim J-w, Gao P, Yustein J (2008) Hypoxia and metabolism—opinion—the interplay between MYC and HIF in cancer. Nat Rev Cancer 8(1):51–56. doi:10.1038/nrc2274

DeBerardinis RJ, Mancuso A, Daikhin E, Nissim I, Yudkoff M, Wehrli S, Thompson CB (2007) Beyond aerobic glycolysis: transformed cells can engage in glutamine metabolism that exceeds the requirement for protein and nucleotide synthesis. Proc Natl Acad Sci U S A 104(49):19345–19350. doi:10.1073/pnas.0709747104

DeBerardinis RJ, Lum JJ, Hatzivassiliou G, Thompson CB (2008) The biology of cancer: metabolic reprogramming fuels cell growth and proliferation. Cell Metab 7(1):11–20. doi:10.1016/j.cmet.2007.10.002

Diaz-Ruiz R, Rigoulet M, Devin A (2011) The Warburg and Crabtree effects: on the origin of cancer cell energy metabolism and of yeast glucose repression. Biochim Biophys Acta 1807(6):568–576. doi:10.1016/j.bbabio.2010.08.010

Gallagher CN, Carpenter KL, Grice P, Howe DJ, Mason A, Timofeev I, Menon DK, Kirkpatrick PJ, Pickard JD, Sutherland GR, Hutchinson PJ (2009) The human brain utilizes lactate via the tricarboxylic acid cycle: a C-13-labelled microdialysis and high-resolution nuclear magnetic resonance study. Brain 132:2839–2849. doi:10.1093/brain/awp202

Galluzzi L, Kepp O, Vander Heiden MG, Kroemer G (2013) Metabolic targets for cancer therapy. Nat Rev Drug Discov 12(11):829–846. doi:10.1038/nrd4145

Garcia-Cao I, Song MS, Hobbs RM, Laurent G, Giorgi C, de Boer VCJ, Anastasiou D, Ito K, Sasaki AT, Rameh L, Carracedo A, Vander Heiden MG, Cantley LC, Pinton P, Haigis MC, Pandolfi PP (2012) Systemic elevation of PTEN induces a tumor-suppressive metabolic state. Cell 149(1):49–62. doi:10.1016/j.cell.2012.02.030

Lee CH, Inoki K, Karbowniczek M, Petroulakis E, Sonenberg N, Henske EP, Guan KL (2007) Constitutive mTOR activation in TSC mutants sensitizes cells to energy starvation and genomic damage via p53. Embo J 26(23):4812–4823. doi:10.1038/sj.emboj.7601900

Matsuno T, Goto I (1992) Glutaminase and glutamine-synthetase activities in human cirrhotic liver and hepatocellular-carcinoma. Cancer Res 52(5):1192–1194

Matsuno T, Hirai H (1989) Glutamine-synthetase and glutaminase activities in various hepatoma-cells. Biochem Int 19(2):219–225

Mazurek S, Boschek CB, Hugo F, Eigenbrodt E (2005) Pyruvate kinase type M2 and its role in tumor growth and spreading. Semin Cancer Biol 15(4):300–308. doi:10.1016/j.semcancer.2005.04.009

Pavlides S, Tsirigos A, Vera I, Flomenberg N, Frank PG, Casimiro MC, Wang CG, Pestell RG, Martinez-Outschoorn UE, Howell A, Sotgia F, Lisanti MP (2010) Transcriptional evidence for the "Reverse Warburg Effect" in human breast cancer tumor stroma and metastasis: similarities with oxidative stress, inflammation, Alzheimer's disease, and "Neuron-Glia Metabolic Coupling". Aging 2(4):185–199

Pierotti MA, Berrino F, Gariboldi M, Melani C, Mogavero A, Negri T, Pasanisi P, Pilotti S (2013) Targeting metabolism for cancer treatment and prevention: metformin, an old drug with multi-faceted effects. Oncogene 32(12):1475–1487. doi:10.1038/onc.2012.181

Shaw RJ, Kosmatka M, Bardeesy N, Hurley RL, Witters LA, DePinho RA, Cantley LC (2004) The tumor suppressor LKB1 kinase directly activates AMP-activated kinase and regulates apoptosis in response to energy stress. Proc Natl Acad Sci U S A 101(10):3329–3335. doi:10.1073/pnas.0308061100

Shim H, Dolde C, Lewis BC, Wu CS, Dang G, Jungmann RA, DallaFavera R, Dang CV (1997) c-Myc transactivation of LDH-A: implications for tumor metabolism and growth. Proc Natl Acad Sci U S A 94(13):6658–6663. doi:10.1073/pnas.94.13.6658

Sonveaux P, Vegran F, Schroeder T, Wergin MC, Verrax J, Rabbani ZN, De Saedeleer CJ, Kennedy KM, Diepart C, Jordan BF, Kelley MJ, Gallez B, Wahl ML, Feron O, Dewhirst MW (2008) Targeting lactate-fueled respiration selectively kills hypoxic tumor cells in mice. J Clin Invest 118(12):3930–3942. doi:10.1172/jci36843

Tong WH, Sourbier C, Kovtunovych G, Jeong SY, Vira M, Ghosh M, Romero VV, Sougrat R, Vaulont S, Viollet B, Kim YS, Lee S, Trepe J, Srinivasan R, Bratslavsky G, Yang YF, Linehan WM, Rouault TA (2011) The glycolytic shift in fumarate-hydratase-deficient kidney cancer lowers AMPK levels, increases anabolic propensities and lowers cellular iron levels. Cancer Cell 20(3):315–327. doi:10.1016/j.ccr.2011.07.018

Vousden KH, Ryan KM (2009) p53 and metabolism. Nat Rev Cancer 9(10):691–700. doi:10.1038/nrc2715

Warburg O (1956) Origin of cancer cells. Science 123(3191):309–314

Warburg O, Wind F, Negelein E (1927) The metabolism of tumors in the body. J Gen Physiol 8(6):519–530. doi:10.1085/jgp.8.6.519

Zhou GC, Myers R, Li Y, Chen YL, Shen XL, Fenyk-Melody J, Wu M, Ventre J, Doebber T, Fujii N, Musi N, Hirshman MF, Goodyear LJ, Moller DE (2001) Role of AMP-activated protein kinase in mechanism of metformin action. J Clin Invest 108(8):1167–1174. doi:10.1172/jci13505

# Chapter 2
# Cancer Hypoxia and the Tumour Microenvironment as Effectors of Cancer Metabolism

Mammalian cells have various control mechanisms that regulate homeostasis, the maintenance of a constant cellular environment. This includes regulating oxygen homeostasis, such that the need for oxygen during oxidative phosphorylation and other metabolic reactions is balanced with the risk of oxidative damage within the cell (Ruan et al. 2009). Hypoxia is the intermediate state between the homeostatic state of normoxia and the complete absence of oxygen in anoxia. Under hypoxia, the survival of a cell, tissue, organ or organism is governed by its ability to detect and respond to oxygen availability and mount an adaptive response facilitating tolerance of the oxygen deprivation.

Within normal tissues, oxygenation levels generally fall within a range of 4–8 %, and hypoxia (below 1–2 % oxygenation) is rare. In tumours however, hypoxia is almost a dominant state. Tumour hypoxia occurs through two routes. Tumour cells are rapidly dividing and grow at a faster rate than the blood vessels supplying them with oxygen. Consequently, many tumour cells reside at distances from vessels that exceed the diffusion distance of oxygen. These cells are said to be in a state of "chronic" or "diffusion-limited" hypoxia. Tumour cells can also be subjected to "acute" or "perfusion-limited" hypoxia. Here cells become hypoxic as a consequence of vascular collapse, which can be transient in nature. Tumour vessels generally differ from those in normal tissues in that they are leaky and immature. Leakiness leads to high interstitial fluid pressure, and immature tumour vessels are much more prone to collapse as a consequence of high interstitial fluid pressure than mature vasculature.

Given that hypoxia will eventually result in cell death, one could query why it matters. The reason is twofold. Hypoxic cells are resistant to many types of therapy (Ruan et al. 2009), and resistant cells can act as foci for tumour relapse. Secondly, exposure to hypoxia causes changes in tumour cells that render them more aggressive and potentially more able to tolerate stress conditions. Given that hypoxia is a fluctuating state within tumours, there develops an almost "selective environment" that favours survival of those cells that can most readily adapt, that coincidentally develop characteristics that confound successful cancer treatment. Indeed, one of the proposed reasons for why the Warburg effect occurs in tumour cells is that it would naturally enable rapid adaptation to oxygen-deprived conditions when the

E. G. Armitage et al., *Correlation-based network analysis of cancer metabolism*, SpringerBriefs in Systems Biology, DOI 10.1007/978-1-4939-0615-4_2,

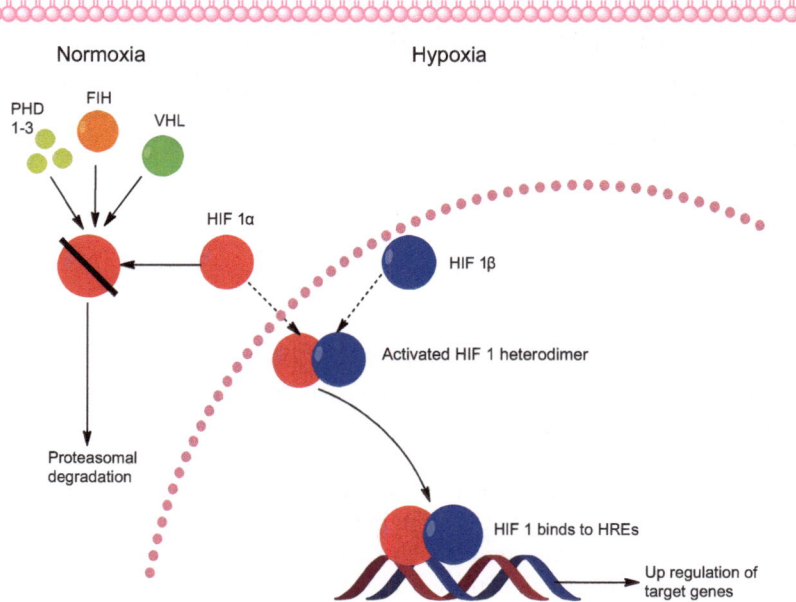

**Fig. 2.1** In normoxia, HIF-1α is hydroxylated by proline PHDs 1, 2 and 3 after which the VHL protein (product of the von Hippel Lindau tumour suppressor gene) is able to tag HIF-1α to be polyubiquitinated for recognition for degradation. In hypoxia, HIF-1α lacks the binding signature for PHDs, VHL and FIH; therefore, HIF-1α and can bind with HIF-1β. This activated heterodimer binds to HREs in the promoter regions of target genes and recruit transcriptional co-activators to enable transcription

cells would be reliant on glycolysis over oxidative phosphorylation for energy production.

## 2.1  Hypoxia-Inducible Factors and Their Role in Tumour Hypoxia

Adaptive response is one of the key mechanisms enabling cell survival under hypoxic conditions. One of the most important pathways orchestrating this response is associated with the activation of the hypoxia-inducible factors (HIFs), in particular HIF-1, which was first described by Wang and Semenza in 1995. The identification of HIF-1 (Wang et al. 1995), its purification (Wang and Semenza 1995) and its molecular characterization (Wang et al. 1995) have all been described previously. The role of HIFs has been well reviewed by Semenza (Semenza 2003, 2012). A schematic for HIF-1 function is depicted in Fig. 2.1. HIF-1 is responsible for regulating the expression of numerous target genes (Semenza 2012). HIF-1 consists of two subunits: HIF-1α and HIF-1β. The HIF-1β subunit is a constitutively expressed nuclear protein, whereas the stability of the HIF-1α subunit is dependent primarily upon oxygen tension (Jaakkola et al. 2001; Lando et al. 2002). HIF-1α is subject to post-translational modification in the presence of oxygen which targets it for degradation (Masson et al. 2001). Oxygen sensors such as prolyl hydroxylases (PHD), factor inhibiting HIF-1 (FIH) and von Hippel Lindau (VHL) protein control the repression of HIF-1α in cells exposed to normal oxygen levels (oxygen partial pressure >10 mmHg). The modification required for targeted degradation is hydroxylation. This is catalysed by the PHDs and occurs on proline residues within the oxygen-dependent degradation domain of HIF-1α (Jaakkola et al. 2001). Hydroxylation at these residues allows for the binding of the VHL tumour suppressor protein which targets HIF-1α for degradation via the proteasome (Jaakkola et al. 2001). In the C-terminal transactivation domain, FIH modifies an asparagine residue, preventing the binding of co-factors necessary for activating the HIF-1 heterodimer (Lando et al. 2002). To activate the heterodimer, HIF-1α translocates to the nucleus to complex with HIF-1β. The heterodimer up-regulates pathways associated with glucose uptake, glycolytic metabolism and pH regulation as well as other features that contribute to the tumour phenotype such as cellular proliferation and differentiation (Troy et al. 2005). The activated HIF-1 complex controls the regulation of genes containing hypoxia response elements (HREs). It does this by interacting with co-factors as well as binding to the promoters of, and facilitating the transcription of, approximately 100–200 genes (Kaelin 2008) that contain HREs (Ruan et al. 2009). HREs are prevalent in genes which encode stress response enzymes, including glycolytic enzymes (Kelly et al. 2008).

It is perhaps unsurprising that metabolism would be a key target of HIF-activation. HIF-1 up-regulates genes encoding glucose transporters (Glut-1 and 3) and nine glycolytic enzymes, including those controlling the rate of glycolytic flux (hexokinase, HK; phosphofructokinase, PFK; and pyruvate kinase, PK) and lactate dehydrogenase (LDH). The shift to anaerobic glycolysis over oxidative phosphorylation is further supported via HIF-1-mediated inhibition of PDH (Semenza 2012). HIF-1 also induces expression of *MCT-4,* which facilitates the metabolic cooperativity between aerobic and hypoxic cells proposed by Sonveaux and colleagues in 2008 (Sonveaux et al. 2008). Here, the hypoxic cells generate and excrete lactate via up-regulated LDH and MCT-1 activity and the aerobic tumour cells utilise the lactate metabolism following uptake via MCT-1 (Sonveaux et al. 2008).

As alluded to earlier, HIF-1 activation is not restricted only to the presence of hypoxia. Both tumour suppressor inactivation and oncogene activation can contribute to HIF activity and have consequential effects on metabolism. At the level of protein stability, loss of VHL (that is frequently observed in renal cancers) prevents HIF-1α degradation under aerobic conditions. Mutations in succinate dehydrogenase (SDH) and fumarate hydratase (FH) lead to increased levels of succinate and fumarate, respectively, which inhibit activity of the HIF–PHD enzymes, resulting in stabilisation in "aerobic" conditions (Isaacs et al. 2005; Pollard et al. 2005; Selak et al. 2005). Loss of phosphatase and tensin homolog (PTEN) can also activate HIF by increasing translation of the HIF-1α subunit (Zundel et al. 2000). In a corollary to the influence of PTEN loss, activation of the phosphpinositide 3-kinase (PI3K)/

Akt/mammalian target of rapamycin (mTOR) axis induces HIF-1 activity, which can be observed in both aerobic and hypoxic conditions. Similarly, RAS/RAF/mitogen-activated protein kinase (MAPK)/extracellular signal-regulated kinase (ERK) (or MAPK/ERK) pathway alterations can favour HIF-activity (Bardos and Athcroft 2004; Shannon et al. 2009; Burrows et al. 2010, 2011). Furthermore cooperativity between HIF-1 and c-myc is observed in the regulation of many targets associated with glycolytic metabolism (Dang et al. 2008).

As for the condition of hypoxia per se, HIF-1 activity can also be associated with poorer response to both radio- and chemotherapy. Mechanisms are complex and given the interplay between HIF/hypoxia and other cancer-associated pathways, it is often challenging to tease out the precise contribution of each separate element to overall therapy response. Indeed, using models with a genetic knockdown of HIF (DN-HIF; see Chap. 5), reveal HIF-dependent and -independent mechanisms of chemotherapy resistance in colorectal tumour cells (Roberts et al. 2009). That said, HIF-1 targeting to improve radio- and chemotherapy response has been ventured as a therapeutic strategy (Meijer et al. 2012; Brown et al. 2006).

Multiple approaches have been proposed for HIF-1 targeting that span blocking HIF transcription, translation, heterodimerisation, DNA-binding activity and HIF-dependent transactivation (Semenza 2012). Deficiencies or blocking of either subunit stops the formation of the active heterodimer and alters the phenotype of the cell in different oxygen potentials, particularly with respect to its metabolic phenotype (Troy et al. 2005). For example, it has been shown that HIF-1$\beta$-deficient cells have an ATP content of up to 80 % lower than corresponding wild type cells (Griffiths et al. 2002). Furthermore, tumour growth is compromised in these models. These studies support that perturbation of metabolism via HIF inhibition can impact on tumourgenecity, but that it is likely through more robust integration of how cells respond to hypoxia and HIF-1 activation will result in additional target identification.

## 2.2 Tumour Hypoxia: The Impact on Treatment

A great deal of research has been directed towards the effects of hypoxia in relation to cancer treatments. This field was driven in the 1950s by Gray, who first confirmed the role of hypoxia in the development of radio-resistance *in vivo* (Bertout et al. 2008). Resistance was shown to be at least three times higher in the anaerobic cells, which became known as the oxygen enhancement effect (Bertout et al. 2008). Radiotherapy relies on oxygen to react with free radicals to generate ionising radiation to damage DNA, leading to cell death (Brown 2000). Hypoxic cancer cells have been reported to be resistant to many anti-cancer chemotherapeutics (Vaupel et al. 2001; Bertout et al. 2008).

The majority of chemotherapy compounds target proliferating cells; however, hypoxic cells have slower proliferation rates (Zolzer and Streffer 2002). Hypoxic cells undergo $G_1$ arrest in the cell cycle resulting in an accumulation of $G_1$ cells (Brown 2000; Amellem and Pettersen 1991). Furthermore, S phase cells exposed to

**Fig. 2.2** Illustration of drug penetration in solid tumours. The rapid growth of the tumour causes poor vasculature. This reduces the diffusion of oxygen as it is metabolised by cells closest the vessel and reduces the penetration of oxygen, nutrients and chemotherapy compounds into areas of the tumour with a poor blood supply

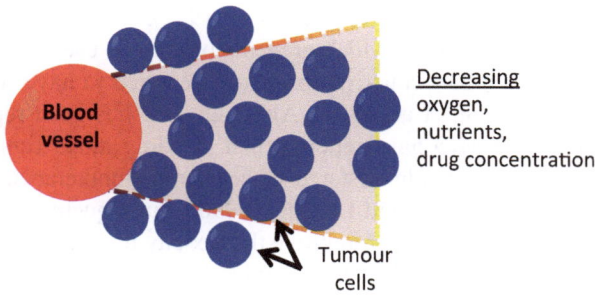

hypoxia arrest progression through the cell cycle following a few hours of treatment (Amellem and Pettersen 1991) suggesting cells in the S phase are more sensitive to hypoxia exposure than the other phases of the cell cycle. Additionally, it has been reported that HIF-1α inhibits *Myc* (an oncogene that drives proliferation in cancer cells) at the molecular level to prevent cell cycle progression (Koshiji et al. 2004).

The vasculature of solid tumours are often found to be insufficiently formed which constrains the transportation of chemotherapeutics into the hypoxic regions of the tumour (Minchinton and Tannock 2006; Fig. 2.2). Cells within close proximity of blood vessels are exposed to the greatest dose of the drug, but hypoxic cells that are located some distance from the blood capillaries receive a lower dose of the chemotherapy drugs due to poor vasculature. Lymphatic deficiency within tumours increases interstitial fluid pressure and changes to the extracellular matrix slow the movement of molecules in the tumour cells (Minchinton and Tannock 2006).

## 2.3   Strategies to Overcome Chemotherapy Resistance in Tumour Hypoxia

Current strategies have been directed towards developing therapeutics that are toxic to hypoxic tumour cells through exploiting their unique properties. One strategy is to improve drug delivery, thus increasing drug accumulation within tumours (Wouters et al. 2002). Erythropoietin, a glycoprotein hormone, is secreted in the kidneys during hypoxia to increase red blood cell production (Wouters et al. 2002). Enhancing haemoglobin levels has shown to improve the cytotoxicity of radiotherapy and chemotherapeutics in hypoxia cancer cells (Wouters et al. 2002). Alternatively, chemotherapies have been developed to specifically target hypoxia cells and are termed hypoxia activated pro-drugs. These are often bio-reductive compounds that are metabolised specifically in low-oxygen environments (Tredan et al. 2007). Killing hypoxic cells, rather than re-oxygenation, may be a smarter method of treatment as this may reduce the occurrence of metastasis (McKeown et al. 1995). Mitomycin C is a chemotherapy agent that uses reductive metabolism to promote toxicity. This mechanism is applicable to hypoxic cells that have limited oxygen available, and it has been shown *in vitro* that the compound selectively kills hypoxic cells compared to normoxic cells (Rockwell et al. 1982). Delayed bone marrow toxicity is a side effect of the drug, which prevents prolonged use of the compound. Tirapazamine (TPZ) is a bio-reductive compound reported to have selective toxicity to hypoxic cells (Zeman et al. 1986). Although TPZ has been reported to be toxic *in vitro,* studies *in vivo* have reported that the drug alone cannot impact tumour growth, and it must be administered with other chemotherapeutics, such as cisplatin, to be effective (Brown and Lemmon 1990). Subsequent studies showed TPZ combined with cisplatin did not improve the toxicity towards advanced head and neck cancer (Rischin et al. 2010). Bis-N-oxide banoxantrone (AQ4N) is a bio-reductive chemotherapy compound, which was developed as an analogue of mitoxantrone (Patterson 1993). Similar to other bio-reductive compounds, AQ4N is reductively activated to AQ4, which binds with high affinity to DNA and acts as a topoisomerase II inhibitor (McKeown et al. 1995). This compound is one of the more successful hypoxia selective drugs developed as it remains active away from the hypoxic region. Furthermore, the drug has been shown to have greater toxicity when applied in combination with other chemotherapeutics, and to date is the most promising compound for treatment of hypoxia-induced chemoresistance (McKeown et al. 1995) Targeting hypoxia remains a focus area for drug development, with several agents currently in clinical trial (Wilson and Hay 2011; Guise et al. 2014).

# References

Amellem O, Pettersen EO (1991) Cell inactivation and cell-cycle inhibition as induced by extreme hypoxia—the possible role of cell-cycle arrest as a protection against hypoxia-induced lethal damage. Cell Prolif 24(2):127–141. doi:10.1111/j.1365-2184.1991.tb01144.x

Bardos JI, Athcroft M (2004) Hypoxia-inducible factor-1 and oncogenic signalling. Bioessays 26(3):262–269. doi:10.1002/bies.20002

Bertout JA, Patel SA, Simon MC (2008) Hypoxia and metabolism series—timeline. The impact of O2 availability on human cancer. Nat Rev Cancer 8(12):967–975. doi:10.1038/nrc2540

Brown JM (2000) Exploiting the hypoxic cancer cell: mechanisms and therapeutic strategies. Mol Med Today 6 (4):157–162. doi:10.1016/s1357-4310(00)01677-4

Brown JM, Lemmon MJ (1990) Potentiation by the hypoxic cytotoxin SR4233 of cell killing produced by fractionated-irradiation of mouse-tumours. Cancer Res 50(24):7745–7749

Brown LM, Cowen RL, Debray C, Eustace A, Erler JT, Sheppard FCD, Parker CA, Stratford IJ, Williams KJ (2006) Reversing hypoxic cell chemoresistance in vitro using genetic and small molecule approaches targeting hypoxia inducible factor-1. Mol Pharmacol 69(2):411–418. doi:10.1124/mol.105.015743

Burrows N, Resch J, Cowen RL, von Wasielewski R, Hoang-Vu C, West CM, Williams KJ, Brabant G (2010) Expression of hypoxia-inducible factor 1 alpha in thyroid carcinomas. Endocr Relat Cancer 17(1):61–72. doi:10.1677/erc-08-0251

Burrows N, Babur M, Resch J, Ridsdale S, Mejin M, Rowling EJ, Brabant G, Williams KJ (2011) GDC-0941 inhibits metastatic characteristics of thyroid carcinomas by targeting both the phosphoinositide-3 kinase (PI3K) and hypoxia-inducible factor-1 alpha (HIF-1 alpha) pathways. J Clin Endocrinol Metab 96(12):E1934–1943. doi:10.1210/jc.2011-1426

Dang CV, Kim J-w, Gao P, Yustein J (2008) Hypoxia and metabolism—opinion—the interplay between MYC and HIF in cancer. Nat Rev Cancer 8(1):51–56. doi:10.1038/nrc2274

Griffiths JR, McSheehy PMJ, Robinson SP, Troy H, Chung YL, Leek RD, Williams KJ, Stratford IJ, Harris AL, Stubbs M (2002) Metabolic changes detected by in vivo magnetic resonance studies of HEPA-1 wild-type tumors and tumors deficient in hypoxia-inducible factor-1 beta (HIF-1 beta): evidence of an anabolic role for the HIF-1 pathway. Cancer Res 62(3):688–695

Guise CP, Mowday AM, Ashoorzadeh A, Yuan R, Lin WH, Wu DH, Smaill JB, Patterson AV, Ding K (Feb 2014) Bioreductive prodrugs as cancer therapeutics: targeting tumor hypoxia. Chin J Cancer 5;33(2):80-86. doi: 10.5732/cjc.012.10285 (Epub 2013 July 12)

Isaacs JS, Jung YJ, Mole DR, Lee S, Torres-Cabala C, Chung YL, Merino M, Trepel J, Zbar B, Toro J, Ratcliffe PJ, Linehan WM, Neckers L (2005) HIF overexpression correlates with biallelic loss of fumarate hydratase in renal cancer: novel role of fumarate in regulation of HIF stability. Cancer Cell 8(2):143–153. doi:10.1016/j.ccr.2005.06.017

Jaakkola P, Mole DR, Tian YM, Wilson MI, Gielbert J, Gaskell SJ, von Kriegsheim A, Hebestreit HF, Mukherji M, Schofield CJ, Maxwell PH, Pugh CW, Ratcliffe PJ (2001) Targeting of HIF-alpha to the von Hippel-Lindau ubiquitylation complex by O-2-regulated prolyl hydroxylation. Science 292(5516):468–472. doi:10.1126/science.1059796

Kaelin WG (2008) The von Hippel-Lindau tumour suppressor protein: O-2 sensing and cancer. Nat Rev Cancer 8(11):865–873. doi:10.1038/nrc2502

Kelly C, Smallbone K, Brady M (2008) Tumour glycolysis: the many faces of HIF. J Theor Biol 254(2):508–513. doi:10.1016/j.jtbi.2008.05.025

Koshiji M, Kageyama Y, Pete EA, Horikawa I, Barrett JC, Huang LE (2004) HIF-1 alpha induces cell cycle arrest by functionally counteracting Myc. Embo Journal 23(9):1949–1956. doi:10.1038/sj.emboj.7600196

Lando D, Peet DJ, Gorman JJ, Whelan DA, Whitelaw ML, Bruick RK (2002) FIH-1 is an asparaginyl hydroxylase enzyme that regulates the transcriptional activity of hypoxia-inducible factor. Genes Dev 16(12):1466–1471. doi:10.1101/gad.991402

Masson N, Willam C, Maxwell PH, Pugh CW, Ratcliffe PJ (2001) Independent function of two destruction domains in hypoxia-inducible factor-alpha chains activated by prolyl hydroxylation. EMBO J 20(18):5197–5206. doi:10.1093/emboj/20.18.5197

McKeown SR, Hejmadi MV, McIntyre IA, McAleer JJA, Patterson LH (1995) AQ4N—an alkylaminoanthraquinone N-oxide showing bioreductive potential and positive interaction with radiation in vivo. Br J Cancer 72(1):76–81. doi:10.1038/bjc.1995.280

Meijer TWH, Kaanders J, Span PN, Bussink J (2012) Targeting hypoxia, HIF-1, and tumor glucose metabolism to improve radiotherapy efficacy. Clin Cancer Res 18(20):5585–5594. doi:10.1158/1078-0432.ccr-12-0858

Minchinton AI, Tannock IF (2006) Drug penetration in solid tumours. Nat Rev Cancer 6(8):583–592. doi:10.1038/nrc1893

Patterson LH (1993) Rationale for the use of aliphatic N-oxides of cytotoxic anthraquinones as prodrug DNA-binding agents: a new class of bioreductive agent. Cancer Metastasis Rev 12(2):119–134. doi:10.1007/bf00689805

Pollard PJ, Briere JJ, Alam NA, Barwell J, Barclay E, Wortham NC, Hunt T, Mitchell M, Olpin S, Moat SJ, Hargreaves IP, Heales SJ, Chung YL, Griffiths JR, Dalgleish A, McGrath JA, Gleeson MJ, Hodgson SV, Poulsom R, Rustin P, Tomlinson IPM (2005) Accumulation of Krebs cycle intermediates and over-expression of HIF1 alpha in tumours which result from germline FH and SDH mutations. Hum Mol Genet 14(15):2231–2239. doi:10.1093/hmg/ddi227

Rischin D, Peters LJ, O'Sullivan B, Giralt J, Fisher R, Yuen K, Trotti A, Bernier J, Bourhis J, Ringash J, Henke M, Kenny L (2010) Tirapazamine, cisplatin, and radiation versus cisplatin and radiation for advanced squamous cell carcinoma of the head and neck (TROG 02.02, HeadSTART): a phase III trial of the Trans-Tasman Radiation Oncology Group. J Clin Oncol 28(18):2989–2995. doi:10.1200/jco.2009.27.4449

Roberts DL, Williams KJ, Cowen RL, Barathova M, Eustace AJ, Brittain-Dissont S, Tilby MJ, Pearson DG, Ottley CJ, Stratford IJ, Dive C (2009) Contribution of HIF-1 and drug penetrance to oxaliplatin resistance in hypoxic colorectal cancer cells. Br J Cancer 101(8):1290–1297. doi:10.1038/sj.bjc.6605311

Rockwell S, Kennedy KA, Sartorelli AC (1982) Mitomycin-C as a prototype bioreductive alkyl-ating agent - invitro studies of metabolism and cyto-toxicity. Int J Radiat Oncol Biol Phys 8(3-4):753–755. doi:10.1016/0360-3016(82)90728-3

Ruan K, Song G, Ouyang GL (2009) Role of Hypoxia in the hallmarks of human cancer. J Cell Biochem 107(6):1053–1062. doi:10.1002/jcb.22214

Selak MA, Armour SM, MacKenzie ED, Boulahbel H, Watson DG, Mansfield KD, Pan Y, Simon MC, Thompson CB, Gottlieb E (2005) Succinate links TCA cycle dysfunction to oncogenesis by inhibiting HIF-alpha prolyl hydroxylase. Cancer Cell 7(1):77–85. doi:10.1016/j.ccr.2004.11.022

Semenza GL (2003) Targeting HIF-1 for cancer therapy. Nat Rev Cancer 3(10):721–732. doi:10.1038/nrc1187

Semenza GL (2012) Hypoxia-inducible factors: mediators of cancer progression and targets for cancer therapy. Trends Pharmacol Sci 33(4):207–214. doi:10.1016/j.tips.2012.01.005

Shannon AM, Telfer BA, Smith PD, Babur M, Logie A, Wilkinson RW, Debray C, Stratford IJ, Williams KJ, Wedge SR (2009) The mitogen-activated protein/extracellular signal-regulated kinase kinase 1/2 inhibitor AZD6244 (ARRY-142886) enhances the radiation responsiveness of lung and colorectal tumor xenografts. Clin Cancer Res 15(21):6619–6629. doi:10.1158/1078-0432.ccr-08-2958

Sonveaux P, Vegran F, Schroeder T, Wergin MC, Verrax J, Rabbani ZN, De Saedeleer CJ, Kennedy KM, Diepart C, Jordan BF, Kelley MJ, Gallez B, Wahl ML, Feron O, Dewhirst MW (2008) Targeting lactate-fueled respiration selectively kills hypoxic tumor cells in mice. J Clin Invest 118(12):3930–3942. doi:10.1172/jci36843

Tredan O, Galmarini CM, Patel K, Tannock IF (2007) Drug resistance and the solid tumor micro-environment. J Natl Cancer Inst 99(19):1441–1454. doi:10.1093/jnci/djm135

Troy H, Chung YL, Mayr M, Ly L, Williams K, Stratford I, Harris A, Griffiths J, Stubbs M (2005) Metabolic profiling of hypoxia-inducible factor-1 beta-deficient and wild type Hepa-1 cells: effects of hypoxia measured by H-1 magnetic resonance spectroscopy. Metabolomics 1(4):293–303. doi:10.1007/s11306-005-0009-8

Vaupel P, Thews O, Hoeckel M (2001) Treatment resistance of solid tumors—role of hypoxia and anemia. Med Oncol 18(4):243–259. doi:10.1385/mo:18:4:243

Wang GL, Semenza GL (1995) Purification and characterization of hypoxia-inducible factor-1. J Biol Chem 270(3):1230–1237

Wang GL, Jiang BH, Rue EA, Semenza GL (1995) Hypoxia-Inducible Factor 1 is a basic-helix-loop-helix-PAS heterodimer regulated by cellular O2 tension. Proc Natl Acad Sci U S A 92(12):5510–5514. doi:10.1073/pnas.92.12.5510

Wouters BG, Weppler SA, Koritzinsky M, Landuyt W, Nuyts S, Theys J, Chiu RK, Lambin P (2002) Hypoxia as a target for combined modality treatments. Eur J Cancer 38 (2):240–257. doi:10.1016/s0959-8049(01)00361-6

Wilson WR, Hay MP (2011) Targeting hypoxia in cancer therapy. Nat Rev Cancer 11(6):393–410. doi:10.1038/nrc3064(Review)

Zeman EM, Brown JM, Lemmon MJ, Hirst VK, Lee WW (1986) SR-4233-a new bioreductive agent with high selective toxicity for hypoxic mammalian-cells. Int J Radiat Oncol Biol Phys 12(7):1239–1242

Zolzer F, Streffer C (2002) Increased radiosensitivity with chronic hypoxia in four human tu-mor cell lines. Int J Radiat Oncol Biol Phys 54 (3):910–920. doi:Pii s0360-3016(02)02963-2. (10.1016/s0360-3016(02)02963–2)

Zundel W, Schindler C, Haas-Kogan D, Koong A, Kaper F, Chen E, Gottschalk AR, Ryan HE, Johnson RS, Jefferson AB, Stokoe D, Giaccia AJ (2000) Loss of PTEN facilitates HIF-1-medi-ated gene expression. Genes Dev 14(4):391–396

# Chapter 3
# Metabolic Fingerprinting of *In Vitro* Cancer Cell Samples

Metabolomics is a commonly used tool in systems biology. Since a range of metabolites can be detected in a single assay, metabolomics can be defined as a holistic and data-driven study of the low molecular weight metabolites present in biological systems (Dunn 2008). The metabolome consists of endogenous and exogenous components: those catabolised or anabolised by the cell or organism itself, or those that are extra-organism or extracellular respectively. The metabolome includes metabolites present in a cell or organism that participate in metabolic reactions required for growth, maintenance and function, as well as metabolites consumed from the external environment. If considering an organism *in vivo,* the external environment could include the metabolomes of interacting organisms, for example from gut microflora in humans (Dunn 2008). In *in vitro* metabolomics (as presented in this research), the external environment is considered the growth medium. Although the functional levels of a biological system include the genome, transcriptome, proteome and metabolomes, the latter is considered most representative of the phenotype (Dunn 2008). Exploring the metabolome following experimental perturbation, where subtle changes can be tractable, may be the best way to reveal the phenotypic changes relative to biological function. For these reasons metabolomics is one of the fastest developing disciplines in systems biology and other aspects of modern science.

Non-targeted metabolomics can be performed by fingerprinting, footprinting or profiling. Although the latter term has been used interchangeably with the former two, it is generally accepted that fingerprinting and footprinting are the truly non-targeted techniques for analysis of the entire metabolome, while profiling focuses on a class of metabolites expected to be associated with a particular biological question under investigation. One could argue that since no analytical technique or combination of analytical techniques currently available is capable of measuring all metabolites that exist in an organism, no experiment can be truly global to satisfy classification of metabolite fingerprinting or footprinting. It is generally accepted that if the intention is to 'blindly' search for a metabolic phenotype in a sample with no prior knowledge about the system then the experimental approach can be defined as fingerprinting or footprinting.

E. G. Armitage et al., *Correlation-based network analysis of cancer metabolism,* 
SpringerBriefs in Systems Biology, DOI 10.1007/978-1-4939-0615-4_3,
© The Authors 2014

Metabolic fingerprinting is a widely used non-targeted approach in metabolomics. Its application spans from comprehensive studies of all detectable metabolites in biological samples to investigate the fate or effect of an exogenous metabolite in an entire system. Although it is not truly quantitative, it is useful for making relative comparisons between biological systems. The presence, absence or relative difference in concentration of the metabolites detected can be compared between experimental groups. These metabolites can be representative of the entire metabolic network and as such the metabolome-wide effects of an environmental or experimental perturbation can be tested. There are many examples of the analysis of *ex vivo* samples including tissue (Sava et al. 2011; Sreekumar et al. 2009) and biofluids (Kind et al. 2007; Dunn et al. 2011; Kenny et al. 2010; Dunn et al. 2008) in mammalian systems. When considering *in vitro* metabolic fingerprinting of mammalian cells, there are many more examples of intracellular fingerprinting rather than extracellular footprinting. There are advantages of footprinting, mainly with respect to the fact that less sample preparation is required, so metabolism can be quenched at a faster rate giving a more representative analysis or a 'snapshot' of metabolism (Kell et al. 2005). However, there is a limit to what metabolites will be present in the footprint. Urine and culture medium for example are largely composed of waste products that are difficult to connect with biological function. It is clearly advantageous to profile extracellular fluids *ex vivo* since sample retrieval is less invasive and more readily available; however for *in vitro* studies fingerprinting the intracellular fingerprint may be more useful in determining properties of biological function. For this reason the metabolic fingerprinting experiments in this research have been based on *in vitro* intracellular fingerprinting, for which there have been several successful protocols developed (Teng et al. 2009; Sellick et al. 2009).

There are some challenges associated with metabolomics that must be considered prior to undertaking research in this area. For example, the volume of metabolites can be too large to analyse and some metabolites cannot be detected through current experimental methods. Additionally, the fluxes and concentrations of metabolites can originate from more than one hierarchical route (controlled by more than one protein, transcription factor or even gene), such that changes observed in the metabolic phenotype of a biological system can be ambiguous with respect to their origin. For full elucidation of the biological system, a combination of the 'omics' can be required. Other challenges in the field are owed to metabolomics being less developed than the preceding 'omic' fields. For example there is a lack of a well-established, comprehensive and publically available database that would be useful in data interpretation and standardisation. In genomics for example, GenBank provides nucleotide sequences for over 380,000 organisms and involves a daily data exchange from laboratories worldwide to continually enhance it (Benson et al. 2012). There have however been advancements towards this for metabolomics, whereby a metabolomics standard initiative (MSI) has been proposed for the identification of metabolites (Fiehn et al. 2007) and laboratory information management systems (LIMS) (Turner and Bolton 2001) including SetupX (Scholz and Fiehn 2007) have been developed. Additionally metabolite libraries have been compiled, an excellent example of which is the Manchester metabolomics database which includes a range of analysed metabolite standards for both gas chromatography mass spectrometry (GC-MS) and ultra-high performance liquid chromatography mass spectrometry (UHPLC-MS); see Brown et al. (2009). Libraries are most advanced for GC-MS in metabolomics, a particular breakthrough for the Agilent Fiehn metabolomics retention time locked library (Kind et al. 2009). Currently the method for identification of features in UHPLC-MS data relies upon accurate mass that can be matched to compounds from web-based sources. A recent advancement in this area has been the development of the Taverna work flow for feature identification (Brown et al. 2011). For other metabolomics platforms such as nuclear magnetic resonance (NMR) spectroscopy, software such as MetaboHunter are available for feature identification (Tulpan et al. 2011). In future it is hoped that the number of features it is possible to identify, will increase and a combined repository for the whole metabolomics community to use will be created that contains data from a wider range of analytical platforms. Some aspects have been addressed by the recently introduced metabolites: a database for comparing metabolomics experiments across species and across analytical platforms (http://www.ebi.ac.uk/metabolights/). The advancements in computational metabolomics so far are enough to make biomarker discovery possible and biomarkers are valuable identifications, regardless of their hierarchical origin, for revealing phenotypic properties in a biological system. In the context of this book, a biomarker can be defined as a representative metabolite of cancer (with respect to experimental treatment) found to be reliably detected in samples of the experimental group or treatment group (Armitage and Barbas 2014). Furthermore, in the identification of key metabolic pathways it is possible that metabolomics alone can reveal potential targets for cancer therapy. Combining the use of different analytical platforms extends the number of metabolites it is possible to detect in biological samples.

The use of quality control in non-targeted metabolomics studies has become routine in recent years. The main purposes of quality control are to pre-condition the instrumental system before analysis of real samples and to ensure quality in the entire analysis by assessing and correcting for analytical variation throughout (signal correction). This is performed by preparing quality control (QC) samples comprised of the range of metabolites included in the samples under investigation. Most commonly, a sample pool is generated from a small volume of each sample for analysis and this pool is prepared in a range of aliquots for use throughout the analysis. In cases where this is not possible for example due to sample volumes being sparse, a synthetic or other substitute QC mix can be used that represents the metabolome of the samples as closely as possible. This may also be useful for large-scale studies where samples are collected and analysed sequentially over a long period of time (Zelena et al. 2009). A key example of the latter (Dunn et al. 2011) described analytical procedures from sample collection and preparation to data acquisition, pre-processing and quality assurance for the analysis of human serum over a period of two years. Included in this report is the suggested use of QC-based robust LOESS signal correction that provides a method both for the integration of multiple analytical batches. This is highly applicable for metabolomics data collected for correlation analysis since datasets are large and even if samples are not collected over a vast period of time, analysis is invariably conducted over a series of analytical batches that must later be aligned for successful data analysis.

Analytical variation in metabolite measurements can be assessed through observing trends in the QC samples. For different analytical platforms, there are suggested thresholds for the relative standard deviation/coefficients of variation deemed acceptable for metabolites in QC samples. For metabolites whose variation lies outside of these thresholds, they are usually removed from the dataset before subsequent analysis. Additionally, if the variation in the QC samples exceeds the variation in the samples for given metabolites, these are also usually removed prior to further analysis. Examples of accepted thresholds are 20% for UHPLC-MS and 30% for GC-MS data. The tolerance is usually greater for GC-MS data that UHPLC-MS data as variation due to chemical derivatisation and injection is higher than variation in UHPLC-MS data (Dunn et al. 2011).

Following pre-alignment of data, there are a range of statistical analyses applied in the field of metabolomics to reveal useful features for biological interpretation of data. Particularly for non-targeted datasets, a combination of univariate and multivariate statistical analysis is usually employed. Depending on the nature of the data (with respect size, variance and normality in data as well as to the number of desired comparisons required for interpretation), it may be more or less appropriate to apply different tests. Commonly used univariate techniques include variations on the student's $t$ test, analysis of variance (ANOVA) and non-parametric tests such as the Wilcoxon sign-ranked and Kruskall–Wallis tests. These tests operate in one dimension, analysing each metabolite in turn comparing experimental groups in a way to reveal trends in individual metabolites with respect to the experimental conditions. Recently, it is becoming an expected requirement that the $p$-values obtained through such analyses are subjected to a multiple testing correction method such as Benjamini-Hochberg, Bonferroni or Storey. These methods for controlling the false discovery rate do so either by correcting the $p$-values or redefining the threshold for significance based on considering the likely number of true positives in the data.

To analyse the co-variation in metabolites with respect to experimental conditions can be considered a useful technique to reveal potential system properties of the metabolome. Popular techniques for this are the unsupervised method of principal components analysis (PCA) and supervised methods such as partial least squares- discriminant analysis (PLS-DA) or principal components—canonical variates analysis (PC-CVA). Metabolomics datasets invariably contain orders of magnitude more variables than samples and sequential analysis and interpretation of each variable can be tedious. This is especially the case for non-targeted analyses where there is no prior expectation of metabolite change, rather the entire dataset is integrated for interesting features of change. Multivariate analysis connects metabolites that change with respect to others, in this way revealing the interesting regions of the metabolome with respect to the experiment. For example, if the metabolomic datasets of two experimental groups are analysed by PCA and observed to separate based on this, the metabolites responsible for the separation can be quickly revealed. Similarly, methods of supervised discriminant function analysis (DFA) can be used to search for patterns in data related to experimental group as pre-defined in the algorithm. For these techniques, detailed descriptions can be referred to elsewhere (Mellinger 1987; Wold et al. 1987; Wold et al. 2001).

Metabolomics has played an important feature of modern cancer research, some of the most notable examples of which have been reviewed (Armitage and Barbas 2014). A key example has combined data with pre-existing gene expression data to determine the mitochondrial glycine biosynthetic pathway to be strongly correlated with rapid proliferation in cancer. This was elucidated from the revelation of the consumption or release of 219 metabolites across a range of cancer cell lines. It was found that rapidly proliferating cells require large quantities of the non-essential amino acid glycine to support growth. Consequently, this study demonstrates the impact of metabolomics studies in identifying new cancer therapeutic targets (Jain et al. 2012). Furthermore, metabolic profiling can be specifically useful for identifying the underlying pathway regulation of metabolic reprogramming (one of the hallmarks of cancer).

# References

Armitage EG, Barbas C (2014) Metabolomics in cancer biomarker discovery: current trends and future perspectives. J Pharm Biomed Anal 87(0):1–11. doi:10.1016/j.jpba.2013.08.041

Benson DA, Karsch-Mizrachi I, Clark K, Lipman DJ, Ostell J, Sayers EW (2012) GenBank. Nucleic Acids Res 40 (D1):D48–53. doi:10.1093/nar/gkr1202

Brown M, Dunn WB, Dobson P, Patel Y, Winder CL, Francis-McIntyre S, Begley P, Carroll K, Broadhurst D, Tseng A, Swainston N, Spasic I, Goodacre R, Kell DB (2009) Mass spectrometry tools and metabolite-specific databases for molecular identification in metabolomics. Analyst 134(7):1322–1332

Brown M, Wedge DC, Goodacre R, Kell DB, Baker PN, Kenny LC, Mamas MA, Neyses L, Dunn WB (2011) Automated workflows for accurate mass-based putative metabolite identification in LC/MS-derived metabolomic datasets. Bioinformatics 27(8):1108–1112. doi:10.1093/bioinformatics/btr079

Dunn WB (2008) Current trends and future requirements for the mass spectrometric investigation of microbial, mammalian and plant metabolomes. Phys Biol 5(1):11001

Dunn WB, Broadhurst D, Ellis DI, Brown M, Halsall A, O'Hagan S, Spasic I, Tseng A, Kell DB (2008) A GC-TOF-MS study of the stability of serum and urine metabolomes during the UK Biobank sample collection and preparation protocols. Int J Epidemiol 37:23–30. doi:10.1093/ije/dym281

Dunn WB, Broadhurst D, Begley P, Zelena E, Francis-McIntyre S, Anderson N, Brown M, Knowles JD, Halsall A, Haselden JN, Nicholls AW, Wilson ID, Kell DB, Goodacre R (2011) Procedures for large-scale metabolic profiling of serum and plasma using gas chromatography and liquid chromatography coupled to mass spectrometry. Nat Protoc 6(7):1060–1083. doi:10.1038/nprot.2011.335

Fiehn O, Robertson D, Griffin J, van der Werf M, Nikolau B, Morrison N, Sumner LW, Goodacre R, Hardy NW, Taylor C, Fostel J, Kristal B, Kaddurah-Daouk R, Mendes P, van Ommen B, Lindon JC, Sansone SA (2007) The metabolomics standards initiative (MSI). Metabolomics 3:175–178. doi:10.1007/s11306-007-0070-6

Jain M, Nilsson R, Sharma S, Madhusudhan N, Kitami T, Souza AL, Kafri R, Kirschner MW, Clish CB, Mootha VK (2012) Metabolite profiling identifies a key role for glycine in rapid cancer cell proliferation. Science 336(6084):1040–1044. doi:10.1126/science.1218595

Kell DB, Brown M, Davey HM, Dunn WB, Spasic I, Oliver SG (2005) Metabolic footprinting and systems biology: the medium is the message. Nat Rev Microbiol 3(7):557–565. doi:10.1038/nrmicro1177

Kenny LC, Broadhurst DI, Dunn W, Brown M, North RA, McCowan L, Roberts C, Cooper GJS, Kell DB, Baker PN (2010) Robust early pregnancy prediction of later preeclampsia using metabolomic biomarkers. Hypertension 56(4):741–749. doi:10.1161/hypertensionaha.110.157297

Kind T, Tolstikov V, Fiehn O, Weiss RH (2007) A comprehensive urinary metabolomic approach for identifying kidney cancer. Anal Biochem 363(2):185–195. doi:10.1016/j.ab.2007.01.028

Kind T, Wohlgemuth G, Lee DY, Lu Y, Palazoglu M, Shahbaz S, Fiehn O (2009) FiehnLib: mass spectral and retention index libraries for metabolomics based on quadrupole and time-of-flight gas chromatography/mass spectrometry. Anal Chem 81(24):10038–10048. doi:10.1021/ac9019522

Mellinger M (1987) Multivariate data-analysis—its methods. Chemom Intell Lab Syst 2 (1-3):29–36. doi:10.1016/0169-7439(87)80083-7

Sava AC, Martinez-Bisbal MC, Van Huffel S, Cerda JM, Sima DM, Celda B (2011) Ex vivo high resolution magic angle spinning metabolic profiles describe intratumoral histopathological tissue properties in adult human gliomas. Magn Reson Med 65(2):320–328. doi:10.1002/mrm.22619

Scholz M, Fiehn O (2007) Setup X—a public study design database for metabolomic projects. Pac Symp Biocomput 12:169–180

Sellick CA, Hansen R, Maqsood AR, Dunn WB, Stephens GM, Goodacre R, Dickson AJ (2009) Effective quenching processes for physiologically valid metabolite profiling of suspension cultured mammalian cells. Anal Chem 81(1):174–183. doi:10.1021/ac8016899

Sreekumar A, Poisson LM, Rajendiran TM, Khan AP, Cao Q, Yu JD, Laxman B, Mehra R, Lonigro RJ, Li Y, Nyati MK, Ahsan A, Kalyana-Sundaram S, Han B, Cao XH, Byun J, Omenn GS, Ghosh D, Pennathur S, Alexander DC, Berger A, Shuster JR, Wei JT, Varambally S, Beecher C, Chinnaiyan AM (2009) Metabolomic profiles delineate potential role for sarcosine in prostate cancer progression. Nature 457(7231):910–914. doi:10.1038/nature07762

Teng Q, Huang W, Collette T, Ekman D, Tan C (2009) A direct cell quenching method for cell-culture based metabolomics. Metabolomics 5(2):199–208

Tulpan D, Leger S, Belliveau L, Culf A, Cuperlovic-Culf M (2011) MetaboHunter: an automatic approach for identification of metabolites from H-1-NMR spectra of complex mixtures. BMC Bioinformatics 12:400. doi:10.1186/1471-2105-12-400

Turner E, Bolton J (2001) Required steps for the validation of a laboratory information management system. Qual Assur (San Diego, Calif) 9(3-4):217–224

Wold S, Esbensen K, Geladi P (1987) Principal component analysis. Chemo Intell Lab Syst 2 (1–3):37–52. doi:10.1016/0169-7439(87)80084-9

Wold S, Sjöström M, Eriksson L (2001) PLS-regression: a basic tool of chemometrics. Chemom Intell Lab Syst 58 (2):109–130. doi:10.1016/S0169-7439(01)00155-1

Zelena E, Dunn WB, Broadhurst D, Francis-McIntyre S, Carroll KM, Begley P, O'Hagan S, Knowles JD, Halsall A, Wilson ID, Kellt DB, Consortium H (2009) Development of a robust and repeatable UPLC-MS method for the long-term metabolomic study of human serum. Anal Chem 81(4):1357–1364. doi:10.1021/ac8019366

# Chapter 4
# Network-Based Correlation Analysis of Metabolic Fingerprinting Data

Correlation analysis, first invented by Francis Galton and later scientifically conceptualised by Karl Pearson, has many powerful applications in biology for describing causality in biological systems. Ever since the 1920s, causation has been connected with correlation in this way. The underlying mechanisms in biological processes are shadowed in correlations that when analysed can reveal connections in biological data that provide a starting point to realise underlying biological processes.

There are different scenarios of sources in variation between biological variables that give rise to the (anti-) correlations between them. In general, they can be defined as specific perturbations, such as the effect of a single gene knockout or enzymatic activity, or as global perturbations, such as the effect of the environment on a biological system or the evolution of the system over time that can simultaneously affect many components in the system (Steuer 2006). One other significant feature of biological systems that give rise to sources of variation is the balance between intrinsic and extrinsic variability. Intrinsic variability is described as the 'probablistic nature of the timing of collision events between reacting biological molecules' (Toni and Tidor 2013). It is an important condition within computational models of biological systems; however, it is not well analysed in the lab. It is associated with noise generated by intrinsic fluctuation and can cause a variable amount of variation in data that can be relevant to correlation analysis. Intrinsic variation is thought to be responsible for some of the most intense interrelations between metabolites that are evident across the population under identical experimental conditions (Steuer 2006). Extrinsic variability is a similar concept but is due to noise arising from outside the boundaries of the system of interest. For example, when considering a small sub-network of metabolism, extrinsic variation could arise from an upstream region of metabolism that is out of the system of interest, but which affects the system in a real situation. Even if one only wants to consider the TCA cycle, it is necessary to consider the off-shoots to this cycle that inherently control its function.

Correlation analysis is a powerful tool to explore cellular phenotype. Cellular phenotype is largely governed by metabolism. There are approximately 2,900 endogenous metabolites that are currently detectable in the human body using analytical techniques such as gas chromatography–mass spectrometry (GC-MS), liquid chromatography–mass spectrometry (LC-MS) or nuclear magnetic reso-

E. G. Armitage et al., *Correlation-based network analysis of cancer metabolism,*
SpringerBriefs in Systems Biology, DOI 10.1007/978-1-4939-0615-4_4,
© The Authors 2014

---

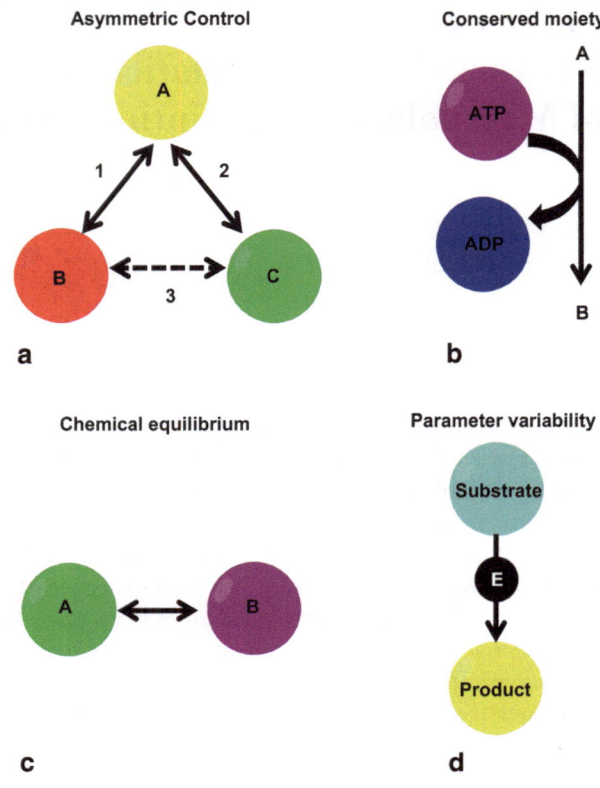

**Fig. 4.1** Examples of factors contributing to the correlation between metabolites. **a** An external factor controls correlation *1* between metabolites *A* and *B* as well as correlation *2* between *A* and *C*. By asymmetric control, this factor also controls a correlation *3* between *B* and *C*. **b** In the energy-consuming metabolic reaction between *A* and *B*, *ADP* and *ATP* are negatively correlated. **c** A correlation between *A* and *B* can occur when both are close to chemical equilibrium. **d** High variability of enzyme *E* controls the negative correlation between its substrate and product

nance (NMR; Wishart et al. 2009). The human metabolic network involves connections between metabolites through biochemical reactions and links many metabolic pathways together into one representation of metabolic function. The structure of this network gives rise to the relative concentrations of metabolites that are present in cells, and naturally, the concentrations of certain metabolites tend to be correlated with others in the network due to their position and influence on other metabolites via key enzymatic reactions. The influence of one metabolite on another can be due to a neighbouring interaction; however, in reality correlated metabolites tend to be spatially or temporally separated in the network. A correlation can exist between two metabolites that could be due to any number of factors. Many of these have been discussed (Camacho et al. 2005). Figure 4.1 shows schematics for some of these factors. Two metabolites can be highly correlated due to the domination of a single parameter whose variability has more control over the correlation than any other parameter (asymmetric control). In such a case, groups of highly correlated metabolites can form if one metabolite is highly correlated to two others due to a single parameter; then the two others by necessity must also be highly correlated and so on. For example, a gene could ultimately control the correlation between metabolites A and B as well as A and C. By asymmetric control, that gene is likely, therefore, to be accountable for a correlation between B and C as well. Other factors that can cause metabolites to be highly correlated include cases where metabolites are highly positively correlated when they are close to chemical equilibrium or highly negatively correlated when they share a conserved moiety. An example of the latter would be the anti-correlation between ATP and ADP such that the concentration of one is higher while the other is lower because phosphate is a conserved moiety transferred cyclically between them. Finally, correlations can be due to high variability in one parameter. For example, high variability in a certain enzyme will pose a negative correlation on its substrates and products.

Different metabolites have different levels of connectivity both in reality and in representations of human metabolic networks. Highly connected metabolites feature in many reactions and can be considered 'hubs' in the network. These hubs may change when cells are exposed to different environments as alternative metabolic pathways are up- and down-regulated to promote survival in that particular environment. Determining hubs and key pathways that change in response to different environments or treatments could give an insight into how cells use metabolism to respond and potentially reveal regions of the network that could be targeted in cancer therapy.

## 4.1   Calculating Correlation Coefficients

There are many different methods for performing correlation analysis. The type and quality of the data to which correlation analysis will be applied along with the level of robustness required usually provide the basis for choosing certain methods over others. The most commonly used are the Pearson's product correlation and Spearman's rank correlation methods. In both, a correlation coefficient ($r$) representative of the connectivity between two independent variables (metabolites) is calculated, ranging from $-1$ to $1$; where a coefficient of $0$ implies no correlation between variables, coefficients in the range $\pm 0.7 - 1$ usually imply strong correlation between variables, and coefficients in the range $\pm 0 - 0.7$ usually imply weak correlation between variables. Pearson's product-moment correlation method computes a coefficient that is invariant to linear transformation in variables (Rodgers and Nicewander 1988), and this type of correlation analysis is only valid when variables are linearly related (Camacho et al. 2005). Further requirements include data that are approximately normally distributed and do not contain outliers. Pearson's correlation is an example of a standard covariance where altering the scaling of the data will affect the variance and covariance.

The following equation shows the calculation used to obtain coefficients for the pairwise correlation analyses between identified metabolites. The Pearson's product-moment correlation equation, where $r$ is the correlation coefficient calculated for the pairwise correlation of variables $x$ and $y$

$$r = \frac{\sum_{i=1}^{n}(x_i - \bar{x})(y_i - \bar{y})}{\left[\sum(x_i - \bar{x})^2 \sum(y_i - y)^2\right]^{1/2}}$$

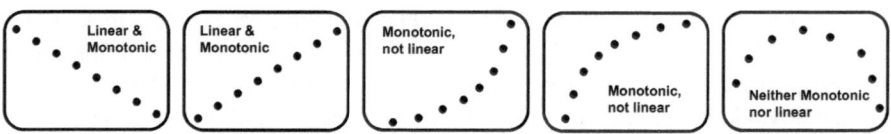

**Fig. 4.2** Patterns for the relationship between two variables that can be linear, and/or monotonic or not

Spearman's correlation coefficient is calculated when two variables are not normally distributed and are non-linearly (but monotonically) related. Although Spearman's rank correlation can be used to analyse non-parametric data and is less sensitive to outliers (Camacho et al. 2005), Pearson's product-moment correlation method is statistically more powerful. The quality and validity of a result from correlation analysis is highly influenced by the sample size. Although Pearson's product-moment correlation method is less sensitive to sample size than Spearman's rank correlation method, in general, correlation analysis should be avoided for experiments with fewer than ten biological replicates (Camacho et al. 2005). For a Spearman's correlation, the data is ranked prior to correlation by its order where the smallest becomes one, the second smallest becomes two and so forth.

The equation for calculating Spearman's rank correlation coefficient ($r_s$), where $n$ represents sample size and $d_i$ is the calculated difference between $x_i$ and $y_i$, is:

$$r_s = 1 - \frac{6\sum d_i^2}{n(n^2 - 1)}$$

Whether the correlation method applied is Pearson's or Spearman's rank method, it is largely governed by whether or not pairwise variables are approximately normally distributed. For both methods, it is assumed that the function describing one variable in terms of the other is monotonic. Such a relationship allows an increase in one variable with an increase in the other, a decrease in one variable with a decrease in the other, or a decrease in one variable with an increase in the other. The patterns that can describe each of these relationships are depicted in Fig. 4.2.

A linear relationship is always monotonic; however, it is also possible to have monotonic relationships between variables that are not linear. In the latter case, it is highly appropriate to apply Spearman's rank method of correlation analysis, since linearity is not a requirement. For Pearson's method, it is necessary to satisfy both linearity and monotonicity. When neither condition is met, correlation analysis must follow more computationally complex approaches that involve form-free regression or permutation testing (Shipley 2004).

*Partial* correlations can be used to distinguish between a direct causal relationship through intermediate variables or directly due to common causes. *Partial* correlations are, therefore, a step towards describing the causal inference. For example, Fig. 4.3 shows the zero-order correlation would calculate A to be correlated to B, B to C and A to C; however, a *partial* correlation would calculate that A is correlated

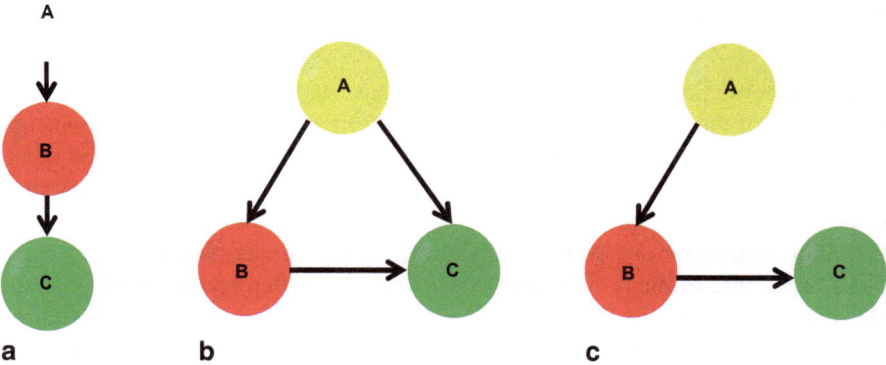

**Fig. 4.3  a** Reaction **b** Full correlation **c** Partial correlation

to B and B is correlated to C as B is the intermediate metabolite and the cause of A and C being correlated.

*Partial* correlation coefficients can be applied similar to zero-order correlations using either Spearman's or Pearson's correlation, as described above, to identify the direct interactions through conditioning each pairwise interaction of two variables against all remaining variables. *Partial* correlation between variables $X$ and $Y$ given $Z$ can be computed using the following equation:

$$\rho_{XY \cdot Z} = \frac{\rho_{XY} - \rho_{XZ}\rho_{ZY}}{\sqrt{1 - \rho^2_{XZ}} \sqrt{1 - \rho^2_{ZY}}}$$

This allows for the identification of correlations among residuals (errors of prediction) where the regression of a variable $X$ on $Y$ when $X$ is subtracted by $X'$ gives the residual $e$. The residual $e$ is not correlated to $Y$, and thus, any correlation $X$ shares with another variable $Z$ cannot be due to $Y$, therefore, direct interactions can be discovered.

Full-order *partial* correlations require the number of samples to be greater than the number of variables. When the number of biological replicates ($n$) is greater than the number of variables ($p$), *partial* correlations can be calculated by inverting the sample covariance matrix. Since this is not the case for most metabolomics and transcriptomics datasets, it is not possible to apply a simple inversion. In order to calculate the partial correlation where the number or variables is greater than the number of samples, the covariance matrix must be estimated by alternative methods (Stifanelli et al. 2011), for example, applying a Moore–Penrose pseudoinverse (PINV).

## 4.1.1   *Correlation Coefficient Transformation*

Data transformations can be applicable at many stages of a statistical analysis. Regardless of the transformation or the stage in the analysis, transformations generically aim to re-mould the data to follow an approximately normal or Gaussian distribution that is necessary for meeting the requirements of many statistical procedures subsequently applied. Furthermore, it is possible to apply transformations that allow non-linear data to be re-moulded to follow a linear trend, making them more applicable to many different statistical analyses that require such relationships. A common non-linear transformation to achieve linearity is the logarithmic transformation. This transformation can be used both to achieve an approximately normal distribution as well as to linearise data. In terms of linearisation, it is applicable when the original data consists of a curved structure that, following logarithmic transformation, can be converted to follow a linear trend. Additionally, many metabolites (and biological variables in general) follow a log normal distribution which means that log transformation corrects the distribution to normal, as required for parametric statistical analyses. This is particularly valuable when data contain outliers at the higher end of the distribution (positive skew) and logarithmic transformation reshapes the data into a Gaussian distribution.

If data are observed to follow a non-normal, non-linear distribution, it is necessary to transform data prior to correlation analysis using Pearson's product moment method; otherwise, it may be more appropriate to apply Spearman's rank correlation analysis. Assuming two metabolites, for which a pairwise correlation coefficient is computed, follow a normal distribution and the relationship between them is linear, it is possible to normalise the correlation coefficient through transformation. These transformations have the aim of variance stabilisation and are used for regression-based and analysis of variance-based statistical methods. These transformations are applicable to data with a bivariate, normal distribution (data that have two independently normally distributed variables; two metabolites).

One such example of variance stabilising type transformation is the Fisher's $\dot{z}$-transformation. This can be applied as an approximation for sample sizes as low as 25, and more exactly, for more than 50 samples where pairwise metabolites are independent. Following this transformation, $\dot{z}$ has an approximately constant variation for all values of correlation. An example of this has recently been published in article (Kotze et al. 2013). Fisher's $\dot{z}$-transformation for normalisation of the correlation coefficient can be computed using the following equation:

$$\dot{z} = \frac{1}{2} ln \frac{1+r}{1-r}$$

Fisher's $\dot{z}$-transformation for correlation analysis is mainly associated with Pearson's product-moment correlation coefficient for bivariate normally-distributed data, however, more generally it can be applied for normalisation of correlation coefficients computed using Spearman's rank method. Other methods for normalisation of correlation coefficients include normalisation to follow a Student's $t$ distribution or the application of a Hotelling transformation. Assuming independence of pairwise metabolites, an exact transformation can be performed so that a correlation coefficient follows the Student's $t$ distribution with $n-2$ degrees of freedom. The following equation is applied for this transformation to follow the Student's $t$ distribution with $n-2$ degrees of freedom:

$$t_r = \frac{r\sqrt{n-2}}{\sqrt{1-r^2}}$$

The Hotelling transformation offers another option for which sample sizes as low as ten can be robustly transformed. The equation for this Hotelling transformation is:

$$z = \sqrt{(n-1)}\left[ 0.5ln\left(\frac{1+r}{1-r}\right) - \frac{1.5ln\left(\frac{1+r}{1-r}\right)+r}{4(n-1)} \right]$$

A more detailed outlook on transformations of correlation coefficients and for testing independence of correlated variables can be found in (Shipley 2004).

## 4.1.2 Determining the Significance of a Correlation Coefficient

Following the determination of $\dot{z}$ using Fisher's $\dot{z}$-transformation described above, it has been proposed that the significance of a correlation coefficient can be tested using the following equation (Fisher 1915):

$$\dot{z}^T = \dot{z}^T\sqrt{N-3}$$

This equation is dependent on sample number ($N$) and is true for data following an approximately Gaussian distribution with 95 % confidence (a significance level of $\alpha = 0.05$) as described in (Fisher 1915). Depending on the level of $\alpha$, it may or may not be appropriate to consider the number of samples required to support this significance. One way to approach this is to consider the number of observations (samples) required for pairwise correlation analysis between variables (metabolites) that satisfies a certain tolerance of standard error. The standard error equation is given by:

$$SE = \frac{(1-r^2)}{\sqrt{n-1}}$$

This equation can be rearranged to determine the sample size that should be used to satisfy both the standard error ($SE$) and correlation coefficient ($r$) that can be chosen for specific purposes. This can be useful to consider in experimental design,

**Table 4.1** Sample sizes required for combinations of standard error and correlation coefficient values calculated using the equation for standard error

| Standard error ($SE$) | Correlation coefficient ($r$) | | | | | | | |
|---|---|---|---|---|---|---|---|---|
| | 0.60 | 0.65 | 0.70 | 0.75 | 0.80 | 0.85 | 0.90 | 0.95 |
| 0.01 | 4097 | 3336 | 2602 | 1915 | 1297 | 771 | 362 | 96 |
| 0.02 | 1025 | 835 | 651 | 480 | 325 | 194 | 91 | 25 |
| 0.03 | 456 | 372 | 290 | 214 | 145 | 87 | 41 | 12 |
| 0.04 | 257 | 209 | 164 | 121 | 82 | 49 | 24 | 7 |
| 0.05 | 165 | 134 | 105 | 78 | 53 | 32 | 15 | 5 |
| 0.06 | 115 | 94 | 73 | 54 | 37 | 22 | 11 | 4 |
| 0.07 | 85 | 69 | 54 | 40 | 27 | 17 | 8 | 3 |
| 0.08 | 65 | 53 | 42 | 31 | 21 | 13 | 7 | 2 |
| 0.09 | 52 | 42 | 33 | 25 | 17 | 11 | 5 | 2 |
| 0.10 | 42 | 34 | 27 | 20 | 14 | 9 | 5 | 2 |
| 0.11 | 35 | 29 | 22 | 17 | 12 | 7 | 4 | 2 |
| 0.12 | 29 | 24 | 19 | 14 | 10 | 6 | 4 | 2 |

since if there is a possibility for flexibility in sample number, a desirable/required standard error and level of correlation for the application can be used to determine the number of observations it is necessary to collect data for. This can be particularly useful in *in vitro* experiments and to some extent all the way through to clinical assays. However, if the sample number is restricted, which is particularly the case in clinical observation, this equation can be used to determine the level of error that will be associated with a given correlation coefficient. Table 4.1 summarises the number of samples required to satisfy certain levels of standard error for different correlation coefficients as calculated using the above standard error equation.

### 4.1.3  Determining the Significance of Difference Between Correlation Coefficients

Once correlation analysis has been performed for pairwise comparisons of all metabolites derived from a fingerprinting study (or other relevant metabolomics study), the data can be used in different ways, many of which are beyond biological interest to find related metabolites in the metabolic network. For example, correlation analysis can be used to prove that two peaks in a chromatogram or spectrum arise from a unique metabolite. This can be particularly interesting for identifying unknown peaks that are not well identified through current methods of database searching. It could be that an unidentified peak is found following correlation analysis to be a less common fragment/derivative/isoform of a metabolite for which other peaks have already been assigned the correct identification. For example, different derivatives of metabolites detected using GC-MS should correlate strongly in a positive direction given their origin is from one unique metabolite. Conversely, it could be useful to cluster a group of unknown peaks (from the same retention time) and by observing the information from each, it is possible to suggest a putative identification based on the differences between them (for example, peaks arising from different fragments of a metabolite).

The significance in the difference between two pairwise correlations can be calculated using the following equation:

$$\hat{z}^{\mathrm{T}} = \frac{|\dot{z}_1 - \dot{z}_2|}{\sqrt{\dfrac{1}{N_1 - 3} + \dfrac{1}{N_2 - 3}}} \text{ with } \dot{z}_i = \frac{1}{2}\log\frac{1+C_i}{1-C_i}$$

As shown in the equation, two Fisher-transformed correlation coefficients can be inputted ($\dot{z}_1$ and $\dot{z}_2$) that have different sample sizes ($N_1$ and $N_2$).

Additionally, a correlation threshold $\hat{z}^{\mathrm{T}}$ can be reported for a sample size $N$. Using the calculation described above can alleviate two restricting cases of correlation analysis. A very low threshold would produce false positives and high thresholds would give false negatives. Correlation coefficients $C_{ij}$ are converted to values $\hat{z}_{ij}$ using an inverse transform:

$$C^T = \frac{e^{(2\dot{z}^T)-1}}{e^{(2\dot{z}^T)+1}} \text{ with } \dot{z}^T = \frac{\hat{z}^T}{\sqrt{N-3}}$$

Consequently, two metabolites are considered significantly correlated when $\hat{z}_{ij} > \hat{z}^{\mathrm{T}}$.

Determining the significance of a difference between correlation coefficients across experimental groups is a powerful way to analyse data. If a matrix of correlation coefficients has been computed for pairwise combinations of all measured variables using data collected from $n$ observations (replicates/samples of the same biological system with equal conditions), it can be compared to a similar matrix of another (different) group of observations. For example, the aim of a study could be to compare the metabolic fingerprints of plasma collected from 30 patients with pancreatic adenocarcinoma to the plasma collected from 30 healthy controls in order to reveal potential metabolic markers of the disease through an untargeted approach. By comparing the difference in each of the correlation coefficients calculated in the same way for each dataset, it is possible to observe statistical changes that could perhaps not have been elucidated either through one-dimensional univariate or even multivariate analysis.

The power of this approach is dependent both on the number of samples per group and the level of correlation. For any sample size, it is possible to ascertain what difference is significant based on the strength of correlation for one experimental group. For a range of sample sizes and a range of correlation coefficients, the difference in correlation coefficients necessary between two groups has been computed and the results are displayed in Table 4.2.

Using the equations described in this chapter, it is possible to calculate correlation coefficients and determine significant differences between correlation coef-

**Table 4.2** Calculated differences necessary for significance between two experimental groups given one correlation coefficient and the number of samples in each group

| Sample size ($N$) | Correlation difference | | | | | | | |
|---|---|---|---|---|---|---|---|---|
| | 0.60 | 0.65 | 0.70 | 0.75 | 0.80 | 0.85 | 0.90 | 0.95 |
| 15 | 0.707 | 0.675 | 0.633 | 0.579 | 0.510 | 0.423 | 0.314 | 0.175 |
| 16 | 0.675 | 0.643 | 0.602 | 0.549 | 0.482 | 0.398 | 0.293 | 0.163 |
| 17 | 0.648 | 0.616 | 0.574 | 0.522 | 0.457 | 0.376 | 0.276 | 0.153 |
| 18 | 0.623 | 0.590 | 0.550 | 0.498 | 0.435 | 0.357 | 0.261 | 0.144 |
| 19 | 0.600 | 0.568 | 0.527 | 0.477 | 0.415 | 0.340 | 0.248 | 0.136 |
| 20 | 0.579 | 0.547 | 0.507 | 0.458 | 0.398 | 0.325 | 0.236 | 0.129 |
| 21 | 0.560 | 0.529 | 0.489 | 0.441 | 0.382 | 0.311 | 0.226 | 0.123 |
| 22 | 0.543 | 0.512 | 0.473 | 0.425 | 0.368 | 0.299 | 0.216 | 0.118 |
| 23 | 0.527 | 0.496 | 0.457 | 0.411 | 0.355 | 0.288 | 0.208 | 0.113 |
| 24 | 0.512 | 0.481 | 0.443 | 0.398 | 0.343 | 0.277 | 0.200 | 0.108 |
| 25 | 0.498 | 0.468 | 0.430 | 0.386 | 0.332 | 0.268 | 0.193 | 0.104 |
| 26 | 0.485 | 0.455 | 0.418 | 0.374 | 0.322 | 0.260 | 0.187 | 0.101 |
| 27 | 0.473 | 0.444 | 0.407 | 0.364 | 0.312 | 0.252 | 0.181 | 0.097 |
| 28 | 0.462 | 0.433 | 0.397 | 0.354 | 0.304 | 0.245 | 0.175 | 0.094 |
| 29 | 0.452 | 0.422 | 0.387 | 0.345 | 0.296 | 0.238 | 0.170 | 0.091 |
| 30 | 0.442 | 0.413 | 0.378 | 0.337 | 0.288 | 0.231 | 0.165 | 0.089 |

ficients in order to compare experimental groups (e.g. disease and control). A full method for this type of analysis in cancer related studies has been published (Kotze et al. 2013).

## 4.2 Network Analysis

Although entire pathways cannot be targeted in cancer therapy, it is useful to identify pathways that link correlated metabolites to determine the enzymes that are responsible for their production and consumption. A gene or transcription factor cannot target metabolites directly. It must target the enzymes that produce or consume the metabolite either directly or via a particular signaling cascade. Identifying metabolites in isolation is useful in revealing metabolic signatures but not in hypothesizing potential targets for therapy. For example, more than 25 enzymes are described in KEGG (Kanehisa and Goto 2000) for their associations with fructose. Without linking fructose to another metabolite, it is not possible to identify pathways to narrow down which enzymes could be used to control the level of fructose.

Correlation analysis between two variables offers a new insight into the relationships in a biological system. These are the result of complex interaction of the biological network; however, the cause of the response remains to be identified. In order to discover their network-based origin, correlations can be mapped onto a human metabolic network. These networks are a reconstruction of the biochemical reactions of human metabolism. There are several genome-scale human metabolic networks freely available for use. The two most popular genome-scale human metabolic models are the global reconstruction of the human metabolic network based on genomic and bibliomic data developed by the Palsson group (HMN-P; Duarte et al. 2007) and the Edinburgh human metabolic network (EHMN) reconstruction (Ma et al. 2007). Both metabolic networks have distinctive features making them more or less valuable for use depending on the investigation. For example, both the EHMN and HMN-P contain a similar number of compounds (approximately 2,700), of which more than half have KEGG references in the EHMN (Ma et al. 2007). The HMN-P has more reactions (approximately 3,800) compared to the EHMN (approximately 2,800); however, the HMN-P contains compartmentalisation such that some reactions are the same as others occurring in different subdivisions of the cell/network and many reactions involve transport of metabolites between compartments. Therefore, the number of different biochemical reactions contained does not differ hugely from the EHMN. Although compartmentalisation provides a more realistic model of metabolism, for this research presented in the case studies (Chaps. 5 and 6), the EHMN was more suitable since it is not compartmentalised. The analysis of cell lysates means that metabolites are no longer localised to compartments of the cell, rather metabolites are analysed as collective pools from all compartments. Therefore, correlations between metabolites are not compartment-specific such that it was not clear which compartments the metabolites originated in. Consequently, using of EHMN allowed, in each case study, a simpler yet more valid network analysis of the correlations was observed. More recently, tissue-specific metabolic models have become available to describe the metabolism of diseases such as breast cancer (Agren et al. 2012) or tissue types such as adipose (Mardinoglu et al. 2013) in addition to a more comprehensive human metabolic model (Thiele et al. 2013). As these models are provided as compartmentalised networks, they would need to be uncompartmentalised prior to use; however, they may be able to discover more relevant metabolic pathways within a dataset taken from a specific tissue or studying a particular disease.

The network must be converted into a stoichiometric network to enable graph theory to be applied to offer an insight into the interactions and connections of metabolites within the network. An example of the construction of a biochemical network into a stoichiometric network of metabolite interaction is shown below (Fig. 4.4). The final matrix is symmetrised to account for directionality of the biochemical reactions. The network is considered to be an undirected graph.

## 4.2.1   Currency Metabolites

Currency metabolites are considered to be highly connected metabolites, typically involved in side reactions, and include metabolites such as water, ATP and ADP. If co-factor metabolites are not removed from the network before any analysis, connections between pairs of metabolites are not always biologically meaningful since currency metabolites are involved in so many reactions and are highly connected in the network (Ma and Zeng 2003). For example, water has 1,083 metabolite connections in the EHMN, and therefore, should be removed prior to network analysis along with other highly connected energy and redox co-factors, including ATP,

$$
A \xrightarrow{R1} B \xrightarrow{R2} C
\qquad
\begin{array}{c}
 \\ A \\ B \\ C
\end{array}
\begin{pmatrix}
R1 & R2 \\
-1 & 0 \\
1 & -1 \\
0 & 1
\end{pmatrix} = S
$$

**a**

$$
\begin{array}{c}
 \\ A \\ B \\ C
\end{array}
\begin{pmatrix}
R1 & R2 \\
0 & 0 \\
1 & 0 \\
0 & 1
\end{pmatrix}
\begin{array}{l} = S^{+} \\ \text{(produced)} \end{array}
\qquad
\begin{array}{c}
 \\ A \\ B \\ C
\end{array}
\begin{pmatrix}
R1 & R2 \\
1 & 0 \\
0 & 1 \\
0 & 0
\end{pmatrix}
\begin{array}{l} = S^{-} \\ \text{(consumed)} \end{array}
$$

**b**

$$
S^{-} \times S^{+T} =
\begin{pmatrix}
1 & 0 \\
0 & 1 \\
0 & 0
\end{pmatrix}
\times
\begin{pmatrix}
0 & 1 & 0 \\
0 & 0 & 0
\end{pmatrix}
=
\begin{array}{c}
 \\ A \\ B \\ C
\end{array}
\begin{pmatrix}
A & B & C \\
0 & 1 & 0 \\
0 & 0 & 1 \\
0 & 0 & 0
\end{pmatrix}
$$

**c**

Fig. 4.4 Progression from a reaction into a stoichiometric network

ADP, AMP, NAD, NADH, NADP, NADPH, CoA, UTP, UDP, UMP, GTP, GDP, $H_2O$, $CO_2$, $O_2$, orthophosphate and hydrogen.

## 4.3 The Use of Network-Based Correlation Analysis in Reality

In these first four chapters, the background of cancer metabolomics (with particular reference to hypoxia in solid tumours), the current topics in metabolic fingerprinting for cancer samples and concepts concerned with correlation analysis and network-mapping of correlated metabolites that explain particular phenotypes have been discussed. In the subsequent two chapters, case studies will be used to exemplify some of these aspects through real experimental applications of network-based correlation analysis associated to tumour hypoxia. In Chap. 5, the application will be used to explore the tumour phenotypes of cells that do and do not express a specific transcription factor known to enable tumour survival in low hypoxic environments. In the second case study (Chap. 6), the same methods have been employed to study mechanisms of chemoresistance that are known to exist in hypoxic tumour environments.

In both case studies, GC-MS metabolic fingerprinting was performed in the same way to create a list of metabolites for correlation analysis. For convenience, the same methods have also been employed for both GC-MS analysis in the lab and correlation analysis. For example, the type of correlation analysis to be applied was Pearson's product-moment correlation analysis. The sample size and acceptable difference between group correlation coefficients for each metabolite were chosen using the tables and equations from this chapter and correlation coefficients were transformed using the Fisher transformation described. It was decided to accept a standard error of 0.1 with a minimum correlation coefficient for one correlation to be 0.7. This meant a minimum requirement of 27 samples per experimental group. To this end, 30 samples were collected to enable room for experimental error. Additionally, given this sample size and coefficient threshold for the first correlation, it was necessary to consider differences in correlation coefficients between groups of at least 0.407 to be deemed statistically significantly different. Finally, the EHMN was chosen as the metabolic network reconstruction onto which pairs of differently correlated metabolites between groups were mapped. This pipeline has been recently published in the accredited peer reviewed scientific journal: BMC Systems Biology (Kotze et al. 2013), and can be consulted for more specific details of the experimental method. With all these parameters set, the following chapters display the power of these methods in unveiling new biological knowledge in these particular fields of cancer research.

# References

Agren R, Bordel S, Mardinoglu A, Pornputtapong N, Nookaew I, Nielsen J (2012) Reconstruction of genome-scale active metabolic networks for 69 human cell types and 16 cancer types using INIT. PLoS Comput Biol 8(5):e1002518. doi:10.1371/journal.pcbi.1002518

Camacho D, de la FA, Mendes P (2005) The origin of correlations in metabolomics data. Metabolomics 1(1):53–63. doi:10.1007/s11306-005-1107-3

Duarte NC, Becker SA, Jamshidi N, Thiele I, Mo ML, Vo TD, Srivas R, Palsson BO (2007) Global reconstruction of the human metabolic network based on genomic and bibliomic data. Proc Natl Acad Sci U S A 104(6):1777–1782. doi:10.1073/pnas0610772104

Fisher RA (1915) Frequency distribution of the values of the correlation coefficient in samples from an indefinitely large population. Biometrika 10(4):507–521

Kanehisa M, Goto S (2000) KEGG: Kyoto Encyclopedia of genes and genomes. Nucleic Acids Res 28(1):27–30. doi:10.1093/nar/28.1.27

Kotze H, Armitage E, Sharkey K, Allwood J, Dunn W, Williams K, Goodacre R (2013) A novel untargeted metabolomics correlation-based network analysis incorporating human metabolic reconstructions. BMC Syst Biol 7(1):107

Ma H, Zeng A-P (2003) Reconstruction of metabolic networks from genome data and analysis of their global structure for various organisms. Bioinformatics 19(2):270–277

Ma HW, Sorokin A, Mazein A, Selkov A, Selkov E, Demin O, Goryanin I (2007) The Edinburgh human metabolic network reconstruction and its functional analysis. Mol Syst Biol 3:135. doi:10.1038/msb4100177

Mardinoglu A, Agren R, Kampf C, Asplund A, Nookaew I, Jacobson P, Walley AJ, Froguel P, Carlsson LM, Uhlen M, Nielsen J (2013) Integration of clinical data with a genome-scale metabolic model of the human adipocyte. Mol Syst Biol 9:649. doi:10.1038/msb.2013.5

Rodgers JL, Nicewander WA (1988) Thirteen ways to look at the correlation coefficient. Am Stat 42(1):59–66

Shipley B (2004) Cause and correlation in biology: a user's guide to path analysis, structural equations and causal inference, 1st edn. Cambridge University Press, Cambridge

Steuer R (2006) Review: on the analysis and interpretation of correlations in metabolomic data. Brief Bioinform 7(2):151–158

Stifanelli P, Creanza T, Anglani R, Liuzzi V, Mukherjee S, Ancona N (2011) A comparative study of Gaussian Graphical Model approaches for genomic data. arXiv preprint arXiv:11070261

Thiele I, Swainston N, Fleming RMT, Hoppe A, Sahoo S, Aurich MK, Haraldsdottir H, Mo ML, Rolfsson O, Stobbe MD, Thorleifsson SG, Agren R, Boelling C, Bordel S, Chavali AK, Dobson P, Dunn WB, Endler L, Hala D, Hucka M, Hull D, Jameson D, Jamshidi N, Jonsson JJ, Juty N, Keating S, Nookaew I, Le Novere N, Malys N, Mazein A, Papin JA, Price ND, Selkov E Sr, Sigurdsson MI, Simeonidis E, Sonnenschein N, Smallbone K, Sorokin A, van Beek JHGM, Weichart D, Goryanin I, Nielsen J, Westerhoff HV, Kell DB, Mendes P, Palsson BO (2013) A community-driven global reconstruction of human metabolism. Nat Biotechnol 31(5):419–425. doi:10.1038/nbt.2488

Toni T, Tidor B (2013) Combined model of intrinsic and extrinsic variability for computational network design with application to synthetic biology. PLoS Comput Biol 9(3):e1002960

Wishart DS, Knox C, Guo AC, Eisner R, Young N, Gautam B, Hau DD, Psychogios N, Dong E, Bouatra S, Mandal R, Sinelnikov I, Xia J, Jia L, Cruz JA, Lim E, Sobsey CA, Shrivastava S, Huang P, Liu P, Fang L, Peng J, Fradette R, Cheng D, Tzur D, Clements M, Lewis A, De Souza A, Zuniga A, Dawe M, Xiong Y, Clive D, Greiner R, Nazyrova A, Shaykhutdinov R, Li L, Vogel HJ, Forsythe I (2009) HMDB: a knowledgebase for the human metabolome. Nucleic Acids Res 37:D603–D610. doi:10.1093/nar/gkn810

# Chapter 5
# Case Study: Systems Biology of HIF Metabolism in Cancer

## 5.1 Hypoxia Inducible Factor—1

Hypoxia and HIFs (particularly the overexpression of HIF-1) are associated with chemotherapy and radiotherapy resistance (Ruan et al. 2009), thus they play a critical role in tumour survival and defence against eradication. Our knowledge and understanding of the mechanisms of HIFs have started to illustrate great scope in the designing and screening of new anticancer therapies (Ruan et al. 2009; Semenza 2012). However, simply developing antagonists to the HIF pathway is not enough as it is not yet established that HIF drives the transformation of a normal cell to a cancer cell (Esteban and Maxwell 2005).

One of the main metabolic targets previously investigated with respect to the role of HIF-1 in cancer hypoxia is glycolysis (Diaz-Ruiz et al. 2009; Troy et al. 2005). Targeting the transcription of genes that code for glucose transporters (such as Glut 1 and Glut 3) (Griffiths et al. 2002), responsible for eliciting downstream changes in a tumour's metabolic phenotype, are just some of its known activities in regulating glycolysis. It is also thought that HIF-1 mediates an adaptation to hypoxia through downregulating the activity of the TCA cycle as well as mitochondrial oxygen consumption through inhibiting PDK1(Kim et al. 2006). The mechanism for this is thought to be an induced expression of both PDK and LDH-A by a hypoxia-driven increase in HIF-1α (Kim et al. 2006). In this way, HIF may be directly responsible for controlling the increased conversion of glucose into lactate in low oxygen microenvironments possessed by cancer cells. In another example, it has been shown that inhibiting the HIF pathway significantly reduces glucose uptake and lactate production *in vitro* while also increasing glutamine uptake (Baker et al. 2012). This highlights the importance of central carbon metabolism as a target of HIF.

Most research has linked HIF-1 with central carbon metabolism. This potentially plays the most vital role in cancer cell metabolism. However, central carbon metabolism has many associated pathways that supply metabolite precursors or produce precursors necessary for many other metabolic processes. Furthermore, there may be unrelated metabolic features of cancer metabolism that appear to be controlled by HIF-1 that are of equal importance in cancer function. This research has aimed to explore the metabolome to reveal metabolic features and may be even new targets

E. G. Armitage et al., *Correlation-based network analysis of cancer metabolism,* 35
SpringerBriefs in Systems Biology, DOI 10.1007/978-1-4939-0615-4_5,

for cancer therapy, with particular reference to inhibiting HIF-1. Due to its applicability in the study of metabolites of central carbon metabolism, as well as its superiority in metabolite identification, gas chromatography-mass spectrometry (GC-MS) was selected as the platform for analysis to demonstrate network-based correlation analysis. A technique with the ability to definitively identify metabolites is more compatible with network analysis since offering several putative identifications for peaks, as in liquid chromatography-mass spectrometry (LC-MS), increases complexity in mapping and reduces the chance of accurate interpretation of the results.

As previously mentioned, in Chap. 3, non-targeted metabolomics can be performed by fingerprinting, footprinting or profiling. Although the latter term has been used interchangeably with the former two in the literature, it is generally accepted that fingerprinting and footprinting are the truly non-targeted techniques for analysis of the entire metabolome, while profiling focuses on a class of metabolites expected to be associated with a particular biological question under investigation. In this case, although from previous evidence it was suspected that HIF-1 would have the greatest effect on central carbon metabolism, the motive was to determine any detectable change in metabolism with no focus on pathway or metabolite class.

Although entire pathways cannot be targeted in cancer therapy, it is useful to identify pathways that link correlated metabolites to determine the enzymes that are responsible for their production and consumption. HIF-1 cannot target metabolites directly, it must target the enzymes that produce or consume the metabolite either directly or via a particular signalling cascade. Identifying metabolites in isolation is useful in revealing metabolic signatures but not in hypothesising potential targets for therapy. Identifying specific pathways offers a way to consider certain enzymes over others in their likelihood as HIF-1 targets and inhibiting these enzymes could be the way to inhibit HIF-1 metabolism.

## 5.2 Network-Based Correlation Analysis of HIF-1 Metabolism

Determining hubs and key pathways that change in response to HIF-1 function or oxygen treatment could provide insight into how cells use metabolism to respond to these stresses and potentially reveal regions of the network that could be targeted in cancer therapy. For this case study, this was achieved by identifying strongly correlated metabolites in HCT 116 cells with normal HIF-1 function wild type (WT), and genetically manipulated counterparts that expressed and dominant negative (DN) variant of HIF-1α that blocks HIF-1 activity (Brown et al. 2006; Roberts et al. 2009). Figure 5.1 depicts the formation and validation of these DN cells.

Cells were exposed to normoxia, hypoxia or anoxia and differences between correlation coefficients due to HIF-1 or oxygen were identified. Once identified, differently correlated metabolites were mapped onto a computational human metabolic network to reveal their network based origins and the connections between them. This offered a systems biology-based approach to study the metabolic effects of hypoxia in cancer as a system rather than by single entities. After all, cancer is

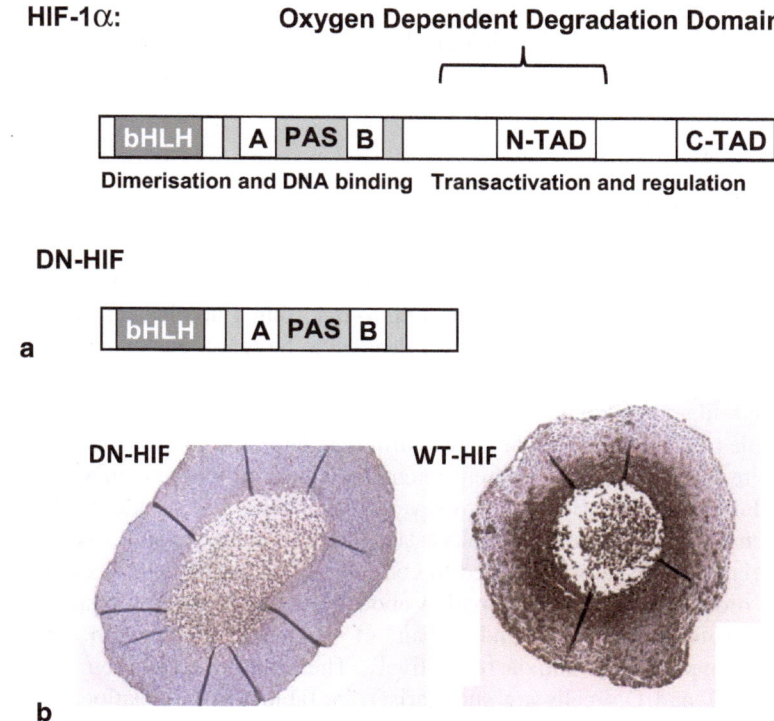

**Fig. 5.1** DN-HIF (**a**; adapted from Brown et al. 2006) is a truncated variant of the HIF-1α that when expressed in cancer cells inhibits HIF-function. HCT 116 cells that express the DN-HIF construct do not induce expression of key HIF-1 targets, including Glut-1 (**b**). HCT 116 DN and WT cells were grown as multicellular spheroids. An oxygen gradient develops from the outer to the inner spheroid region, which induces Glut-1 expression (*brown* staining) in WT but not DN-HIF spheroids. Key: *bHLH* basic helix loop helix, *PAS* per ARNT sim, *TAD* transactivation domain

a multifactorial disease and should be studied from a system perspective. Just as cancer is not controlled by a single gene, protein or metabolic pathway, it was expected that there are many correlated metabolites and that a combination of these are responsible for the cancer phenotype.

Correlated metabolites were identified within the Edinburgh Human Metabolic Network (EHMN) and the shortest path between these was computed from the reactions available in the network. This allowed the connection between correlated metabolites to be observed. From a list of differently correlated metabolites between two experimental groups at a time it was possible to collect pathways between them and together create new sub-networks to describe the network based origin of the differences. For example, the differences between correlations for normoxia and hypoxia show the changes in pathway regulation in response to hypoxia.

The metabolic fingerprinting data obtained from GC-MS analysis of HCT 116 WT and DN cell extracts had an approximately normal distribution after log transformation and any observed metabolite outliers were replaced by a mean peak area from the respective experimental group. The relationships between metabolites were linear in nature and the biological replicate size was moderate ($n=30$) for each experimental group. The data were therefore deemed suitable for Pearson's product-moment correlation analysis.

## 5.2.1   Correlation Analysis

A total of 42 metabolite peaks were identified in HCT 116 WT and DN cells exposed to normoxia, hypoxia or anoxia according to reporting guidelines as described by the MSI (Sumner et al. 2007) that were deemed to be reproducible as measured in the QC samples. Of these, peaks were either definitively identified as single metabolites while others could not be. Additionally, some identification was assigned to multiple peaks. This was either due to different derivatisation products of a single metabolite that elutes with different retention indices or where metabolites could not be definitively identified due to several metabolites having identical electron ionisation mass spectra such as peaks assigned allose/mannose/galactose/glucose.

Correlation analysis was applied to compare WT and DN cells at each oxygen level in order to elucidate information about HIF-1 activity with respect to metabolism. This yielded 16, 42 and 24 pairs of differently correlated metabolites in normoxia, hypoxia and anoxia respectively. The greatest correlation differences between WT and DN cells are summarised in Table 5.1. Correlations are listed according to the oxygen condition they refer to in descending order of correlation difference. The Pearson's product-moment correlation coefficient is given for each.

Significant differences were found between WT and DN cells across a broad spectrum of metabolites and so it could not be concluded that HIF-1 affects any one specific region of metabolism; rather its effect on cells exposed to different oxygen could be metabolome wide. The metabolites that were most correlated in WT samples, irrespective of the difference in correlation between WT and DN included xylitol/ribitol strongly correlated to aspartate, methionine, norleucine, *scyllo*/*myo*-inositol and tyramine/tyrosine as well as malate correlated to glycerol and tyramine/tyrosine. There are no reported chemical reactions involving ribitol in humans and therefore, that peak likely derived from xylitol. Xylitol is involved in the pentose and glucuronate interconversions pathway and can be converted in humans to D-xylose, which is used in starch and sucrose metabolism and can be converted to and from many other sugars including glucose and fructose. The correlation between xylitol and many other metabolites could be indicative of it being used as a carbon source to fuel many processes under low oxygen stress.

Malate is a key player in central carbon metabolism that is interconverted to fumarate which feeds into tyrosine metabolism (involving tyrosine and tyramine). It is likely that this is the structure behind the correlation between malate and tyrosine/tyramine with HIF-1 regulating both the TCA cycle and tyrosine metabolism together. Succinate can also feed into tyrosine metabolism and could also be involved in this mechanism.

**Table 5.1** Pair-wise correlations with a difference greater than 0.5. For each pair of metabolites, the Pearson's product-moment correlation coefficients in HCT 116 WT and DN cells are given along with the difference in correlation between them. The table is split into correlation differences observed at each oxygen level: (a) normoxia (21 % oxygen), (b) hypoxia (1 % oxygen) and (c) anoxia (0 % oxygen)

| Metabolite A | Metabolite B | Difference | $r$ (WT) | $r$ (DN) |
|---|---|---|---|---|
| Normoxia (21 % oxygen) | | | | |
| Allose/Mannose/Galactose/Glucose | 2-oxoglutarate | 0.708 | 0.792 | 0.084 |
| Allose/Mannose/Galactose/Glucose | Threonine | 0.679 | 0.767 | 0.088 |
| Fructose/Sorbose | 2-oxoglutarate | 0.638 | 0.779 | 0.142 |
| Malate | Threonine | 0.595 | 0.718 | 0.122 |
| Glutamate | Malate | 0.536 | 0.785 | 0.249 |
| 2-oxoglutarate | 5-oxoproline | 0.526 | 0.836 | 0.309 |
| Hypoxia (1 % oxygen) | | | | |
| Tyramine/Tyrosine | Xylitol/Ribitol | 0.845 | 0.918 | 0.073 |
| Xylitol/Ribitol | Erythronate/threonate | 0.811 | 0.875 | 0.064 |
| Xylitol/Ribitol | Pyruvate | 0.719 | 0.837 | 0.118 |
| Xylitol/Ribitol | 4-hydroxyproline | 0.693 | 0.834 | 0.141 |
| Fructose | Xylitol/Ribitol | 0.687 | 0.780 | 0.093 |
| Allose/Mannose/Galactose/Glucose | Xylitol/Ribitol | 0.656 | 0.845 | 0.189 |
| Scyllo/Myo-inositol | Xylitol/Ribitol | 0.641 | 0.966 | 0.325 |
| Allose/Mannose/Galactose/Glucose | Glutamine | 0.636 | 0.866 | 0.230 |
| Xylitol/Ribitol | Malate | 0.635 | 0.704 | 0.069 |
| Methionine | Xylitol/Ribitol | 0.614 | 0.973 | 0.359 |
| Xylitol/Ribitol | Aspartate | 0.601 | 0.929 | 0.329 |
| Allose/Mannose/Galactose/Glucose | Erythronate/Threonate | 0.594 | 0.924 | 0.329 |
| Tyramine/Tyrosine | Fructose/Sorbose | 0.594 | 0.905 | 0.312 |
| Xylitol/Ribitol | Norleucine | 0.584 | 0.934 | 0.350 |
| Fructose/Sorbose | Glutamine | 0.578 | 0.866 | 0.288 |
| Creatinine | Xylitol/Ribitol | 0.575 | 0.749 | 0.174 |
| Glutamate | Xylitol/Ribitol | 0.562 | 0.741 | 0.179 |
| Allose/Mannose/Galactose/Glucose | Glycerol | 0.534 | 0.800 | 0.266 |
| Anoxia (0 % oxygen) | | | | |
| Glutamate | Beta-alanine | 0.867 | 0.858 | −0.008 |
| Scyllo/Myo-inositol | Malate | 0.769 | 0.805 | 0.036 |
| Hypotaurine | Putrescine | 0.721 | 0.752 | 0.030 |
| Hypotaurine | Glutamate | 0.698 | 0.822 | 0.124 |
| Methionine | Lactate | 0.682 | 0.773 | 0.091 |
| 5-oxoproline | Malate | 0.681 | 0.833 | 0.152 |
| Tyramine/Tyrosine | Malate | 0.657 | 0.911 | 0.254 |
| Scyllo/Myo-inositol | Beta-alanine | 0.653 | 0.707 | 0.054 |
| Malate | Norleucine | 0.650 | 0.870 | 0.219 |
| Xylitol/Ribitol | Malate | 0.613 | 0.856 | 0.244 |
| Methionine | Malate | 0.607 | 0.853 | 0.247 |
| Malate | Glycerol | 0.591 | 0.902 | 0.311 |
| Xylitol/Ribitol | Lactate | 0.587 | 0.789 | 0.202 |
| Allose/Mannose/Galactose/Glucose | 4-hydroxyproline | 0.552 | 0.832 | 0.281 |
| Allose/Mannose/Galactose/Glucose | Lactate | 0.541 | 0.758 | 0.217 |

From the table it can be seen that correlation differences were due to metabolites being correlated in WT cells and the correlation being lost in HIF-1 deficient DN cells. These connections between metabolites potentially provide the structure behind cancer cell survival through HIF-1 mediated processes. There were some significant differences in correlations where metabolites were correlated in DN cells that were not correlated in WT cells but none exceeding a difference of 0.5.

Although there were fewer differences in normoxia than in the lower oxygen conditions, it was expected that WT and DN cells should have behaved the same under normoxic conditions and that there should be no significantly differently correlated metabolites. Additionally, these differences were not consistent between oxygen conditions so they could not be considered as artefacts of the cell lines. Rather, correlation analysis may be sensitive to very subtle differences caused by HIF-1 in each oxygen condition including normoxia that are not observed in other data analyses.

The best way to consider the response of cells with and without HIF-1 is to determine differently correlated metabolites as a response to oxygen level change. This would reveal potential mechanisms for how HIF-1 promotes metabolic changes in response to low oxygen or which mechanisms DN cells use in the absence of HIF-1 to promote survival in low oxygen. Therefore, a comparison between normoxic WT cells and hypoxic WT cells in addition to comparing normoxic WT cells with anoxic WT cells has been made. A difference in correlation was significantly greater than 0.407 (as determined using the tables in Sect. 5.1.3). There were 22 correlations exhibiting a difference greater than 0.407 for WT normoxia vs. hypoxia and 12 correlations exhibiting a difference greater than 0.407 for WT normoxia vs. anoxia. The same correlation analysis was applied to DN cells to elucidate the greatest differences caused by decreasing oxygen availability to the cells, but that could not be due to HIF-1. In this case there were 22 and 20 correlations meeting the requirements for WT normoxia vs. hypoxia and normoxia vs. anoxia respectively. Although there were 22 differences in pair-wise correlations between normoxia and hypoxia in both WT and DN cells, these pairs were not common, mainly due to the difference in metabolic profiles observed in hypoxia.

The greatest difference in correlation between WT cells exposed to normoxia and WT cells exposed to hypoxia occurred in the correlation between log transformed GC peaks identified as 4-hydroxyproline and allose/mannose/galactose/glucose. The difference was calculated to be 0.804 and was also observed to be the largest difference in correlation between WT cells exposed to normoxia and WT cells exposed to anoxia where the difference was 0.805. The GC peak areas observed for each metabolite in each normoxia and anoxia sample are plotted in Fig. 5.2.

When comparing DN cells exposed to normoxia to cells exposed to hypoxia, the greatest difference in correlation was between citrate and malate which were positively correlated in hypoxia ($r=0.708$) but weakly negatively correlated in normoxia ($r=-0.224$). In anoxic conditions the greatest difference in correlation was between citrate and aspartate which were correlated in anoxia ($r=0.9$) but not in normoxia ($r=0.137$). The log transformed GC peak areas observed for citrate vs. malate in normoxia and hypoxia samples are plotted in Fig. 5.3. From this it can be clearly seen that the two metabolites were strongly correlated in hypoxic cells (blue) but a very weak negative correlation occurred between the metabolites in normoxic cells.

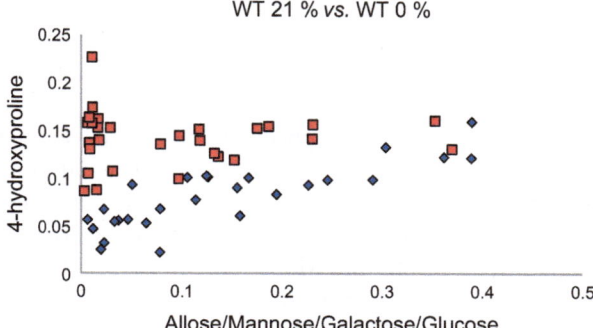

**Fig. 5.2** Pearson's product-moment correlation between 4-hydroxyproline and allose/mannose/galactose/glucose in HCT 116 wild type (WT) cells exposed to normoxia (21 %) represented in *red* squares or anoxia (0 %) represented in *blue* diamonds. The difference in correlation was calculated to be 0.805, where the correlation coefficients were 0.027 and 0.832 for WT normoxia and WT anoxia samples respectively

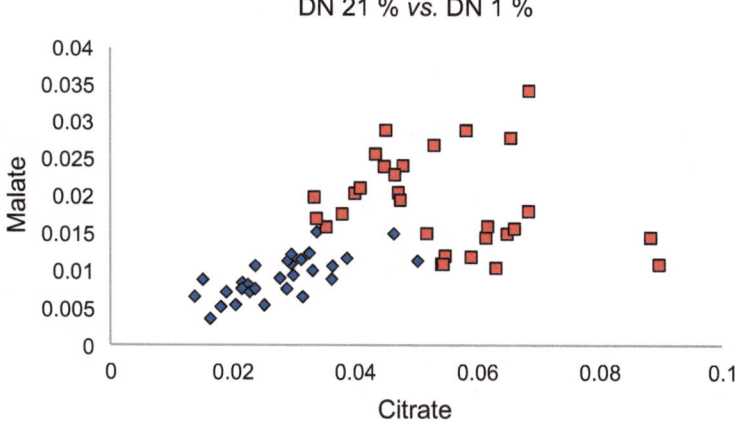

**Fig. 5.3** Pearson's product-moment correlation between citrate and malate in HCT 116 dominant negative (DN) cells exposed to normoxia (21 %) represented by red squares or hypoxia (1 %) represented by blue diamonds. The difference in correlation was calculated to be 0.931, where the correlation coefficients were 0.224 and 0.708 for DN normoxia and DN anoxia samples respectively

## 5.2.2 Network-Based Correlation Analysis

While correlation analysis can be useful to identify key metabolites or phenotypic "hubs", the information gained from correlation analysis alone is vast and to some extent ambiguous. Network-based correlation analysis can therefore be extended to study the data in terms of systems properties. To exemplify this, correlated metabolites from the previous analysis have been mapped onto the EHMN to reveal the shortest pathway between them which could be vital in promoting cancer cell survival under certain environmental conditions such as hypoxia.

This makes possible the construction of new sub-networks of these pathways that can be useful in distinguishing cross over in metabolic pathways and potentially reveal metabolic hubs that are not directly correlated but exist in many pathways connecting correlated metabolites.

Where possible correlations were mapped onto the EHMN and the shortest path between each pair of metabolites in the model was calculated. In some cases singly identified metabolites from the GC-MS data correspond to multiple EHMN metabolites. For example, aspartate corresponded to either L-aspartate or D-aspartate. Since there was no way of identifying which form of this metabolite was detected in the cells, correlations were mapped using both options. It was decided not to map correlations involving the GC peak identified as allose/mannose/galactose/glucose as there were too many options and therefore such an exercise would not have enabled further understanding of the system. In some cases where GC peaks were identified to two metabolites, only one existed in the EHMN. In these cases the metabolite present in the EHMN was used. For example *scyllo*-inositol was not in the EHMN but *myo*-inositol was. Mapping correlations involving xylitol/ribitol was not successful since ribitol was not in the EHMN and xylitol was not highly connected in the network and therefore these correlations have not been represented in sub-network reconstructions. Similarly, fructose/sorbose was mapped as fructose since sorbose was connected only with one other metabolite in the network and did not link up to any of its correlation-paired metabolites.

Correlations that were not mapped due to metabolites not being present in the model, or there being too many options in the model to sensibly assign pathways, can limit the study to a certain extent and therefore it is important to not simply reject these correlations but use the information in addition to the network reconstructions. For example the xylitol/ribitol peak was highly correlated in hypoxic WT cells and these correlations were lost with the absence of HIF-1 in DN cells. More than half of the differences in correlations between WT and DN cells in hypoxia involved this peak. The greatest difference between hypoxic WT and DN cells that involved the correlation between tyramine/tyrosine and xylitol/ribitol which were correlated in WT cells ($r=0.918$) represented in red but not in DN cells ($r=0.073$) represented in blue. This correlation is presented in Fig. 5.4.

Moreover, it is important to highlight that some assumptions must be made that may or may not be biologically correct, for example when mapping correlations concerning the *scyllo*/*myo*-inositol peak as just *myo*-inositol. Nevertheless many correlations could be mapped onto the network and pathways visualised for different sub-networks as discussed below. Using this technique it is possible to visualise inter-connecting pathways regulated by each cell type under each oxygen tension and identify similarities and differences between sub-networks with respect to the pathways involved and the metabolites that appear to be "hubs".

Network analysis was first applied to correlations gained in DN cells that were identified during the comparison between WT and DN samples at each oxygen level. This was done to assess how cells coped with a deficiency in HIF-1 at each oxygen level in isolation. Pathways were identified that could be involved in cancer cell survival over the range of oxygen potentials in the absence of HIF-1 and are shown in Fig. 5.5.

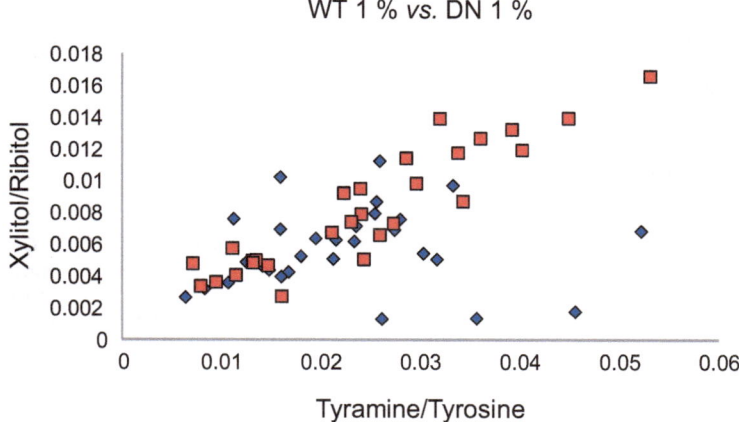

**Fig. 5.4** Pearson's product-moment correlation between tyramine/tyrosine and xylitol/ribitol in HCT 116 cells exposed to hypoxia (1%). The log transformed gas chromatography (GC) peak areas are plotted for these two metabolites where WT samples are shown in *red* squares and DN samples in *blue* diamonds. The difference in correlation was calculated to be 0.845, where the correlation coefficients were 0.918 and 0.073 for WT and DN samples, respectively

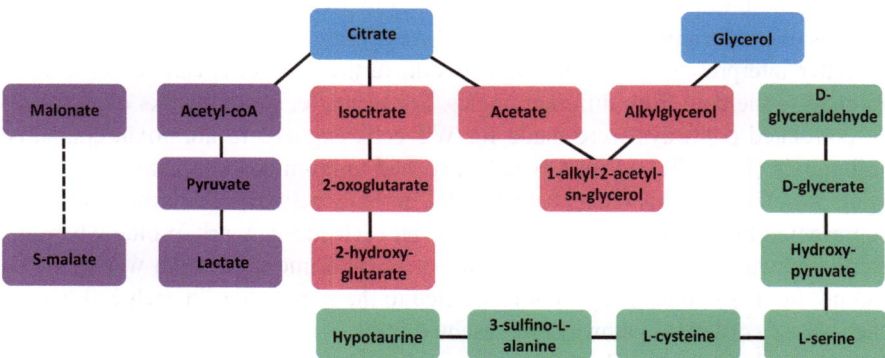

**Fig. 5.5** A schematic of all correlations that were identified in HCT 116 DN samples that were significantly different to WT samples. Metabolite nodes coloured in *turquoise* were normoxic, *lilac* were hypoxic, *red* were anoxic and *white* were shared between oxygen levels. The *dotted line* connecting malonate to (s)-malate is the only correlation that could not be mapped onto the EHMN due to malonate not being in the model

Using this network, revealed pathways should be studied to interpret the analysis with respect to the biological system. An example of this approach is as follows: The pathway connecting hypotaurine to glycerol in normoxic DN cells involved three main pathways as described in KEGG (Kanehisa and Goto 2000): taurine and hypotaurine metabolism (ko00430) for hypotaurine through to L-cysteine, connected to D-glycerate via glycine, serine and threonine metabolism (ko00260) and glycerolipid metabolism (ko00561) to connect D-glycerate through to glycerol.

**Fig. 5.6** Hypotaurine and glycerol are connected via 3 main Kyoto encyclopedias of genes and genomes (KEGG) pathways: taurine and hypotaurine metabolism, glycine serine and threonine metabolism and glycerolipid metabolism

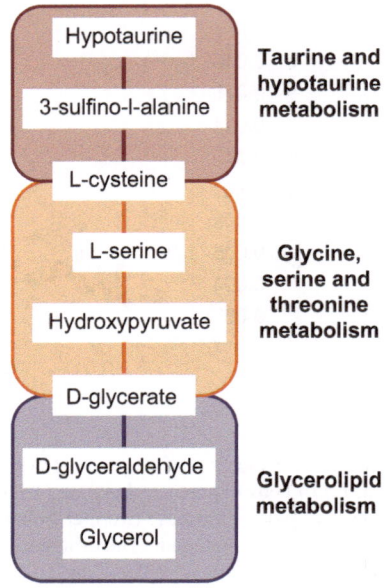

Figure 5.6 shows how the identified pathway between hypotaurine and glycerol crosses over 3 "traditional" metabolic pathways represented by KEGG.

After interpreting the initial sub-network, further networks can be constructed to analyse the data from different angles. For example, sub-networks of differently correlated pathways were made for WT cells exposed to normoxia compared to hypoxia (Fig. 5.7) and for DN cells exposed to normoxia compared to hypoxia (Fig. 5.8). Cross comparisons between WT and DN networks were then made and represented by colouring metabolite nodes on each sub-network in blue where the same pathway was regulated or yellow where the same metabolite was involved but not in the same pathway(s) or connected to the same nodes in each cell line. In the case of conserved pathways represented in blue, such metabolites are likely to be regulated by non-HIF-1 mediated responses to hypoxia. Pathways unique to WT cells were assumed to be regulated by HIF-1 and pathways unique to DN cells were assumed to be regulated as alternative coping mechanisms in the absence of HIF-1. The advantages of assessing these sub-networks are to reveal the HIF-1 specific mechanisms that could be targeted to reduce HIF-1 mediated survival in cancer cells (orange nodes), to identify the alternative metabolic routes HIF-1 deficient cells use that could be employed by WT cells if HIF-1 pathways are truncated in therapy (grey nodes) and the pathways that are common (HIF-1 independent) and appear to be central in cancer cell metabolism and perhaps the best target in cancer therapy.

When considering the common pathways between WT and DN cells in normoxia compared to hypoxia, it appeared that the response was largely centred on citrate with its involvement in the TCA cycle and its connection to *myo*-inositol. Some of these connections were observed to be more significant in DN cells when directly comparing hypoxic and anoxic WT and DN cells (Fig. 5.5), but they were also

**Fig. 5.7** Sub-network representation of pathways between metabolites that were shown to be differently correlated between normoxic and hypoxic HCT 116 WT samples. Metabolite nodes coloured *orange* were unique to WT cells, those coloured in *blue* were also observed in the comparison between normoxic and hypoxic DN samples and *yellow* nodes are metabolites that were also observed in the DN comparison but that were not connected to the same neighbours as they were in WT samples

relevant features of WT metabolism when assessing the metabolic changes associated from the shift from a normoxic to hypoxic environment. Furthermore, this network included more connections that were missed by simply comparing WT and DN cells at each oxygen environment in isolation. Central carbon metabolism controls to some extent most other regions of metabolism through the energy and

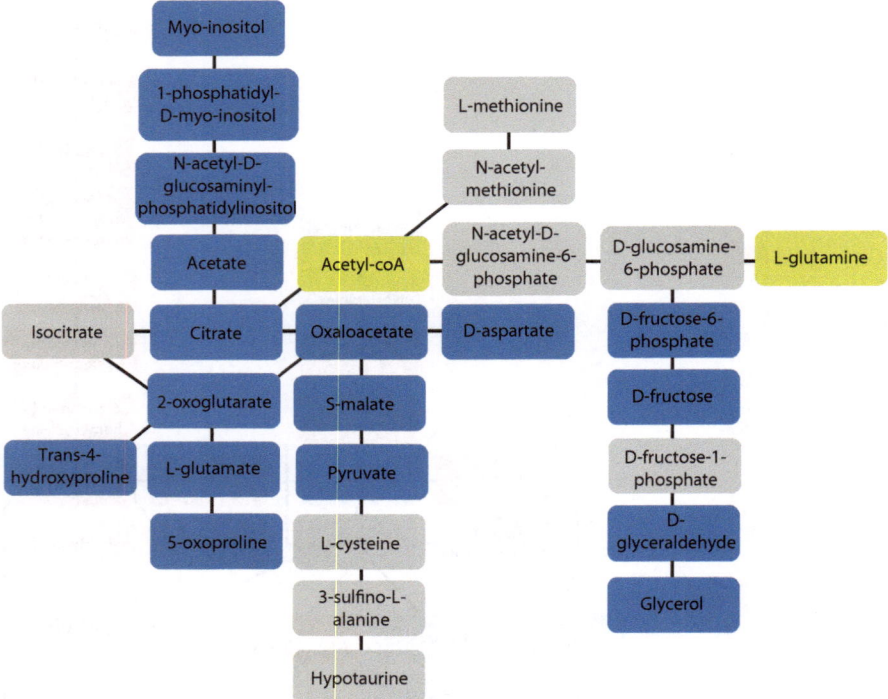

**Fig. 5.8** Sub-network representation of pathways between metabolites that were shown to be differently correlated between normoxic and hypoxic HCT 116 DN samples. Metabolite nodes coloured *grey* were unique to DN cells, those coloured in *blue* were also observed in the comparison between normoxic and hypoxic WT samples and *yellow* nodes are metabolites that are also observed in the WT comparison but that were not connected to the same neighbours as they were in DN samples

bio-precursors generated. Carbon utilisation in nucleic acid, polyamine and amino acid metabolism that all sprouts from central carbon metabolism are known targets of the proto-oncogene c-Myc as a response to hypoxia (Gordan et al. 2007).

## 5.3  Conclusion

Network-based correlation analysis of metabolites measured using GC-MS has proved a novel and highly useful tool to visualise the responses of HCT 116 cells to low oxygen when compared to normoxia as a control. Using this method many correlations were identified including those known to be associated with colon carcinomas; for example *myo*-inositol was a common node irrespective of HIF-1 function and is known to be involved in volume and osmo-regulation that is particularly important in colon carcinomas (Griffin and Shockcor 2004). Furthermore, those associated with low oxygen irrespective of HIF-1 function and those that are specifically HIF-1 mediated were also discovered. Pathways have been identified in each scenario, highlighting regions of the metabolome that could be targeted in cancer therapy, in particular colon carcinoma therapy in the future.

Pathways are often compounded of different features within "traditional" pathways although they have provided an alternative way of viewing cancer metabolism. Analysing sub-networks showing the change in pathway regulation caused by lowering the oxygen microenvironment cells were exposed to have enabled a clearer understanding of the metabolic effects of HIF-1 and hypoxia in general. Additionally, it has revealed alternative pathways that can mediate cancer cell survival in low oxygen environments if HIF-1 pathways were to be targeted. Truncating HIF-1 metabolic pathways will likely induce an up-regulation of the responses observed in DN cell metabolic profiles. This could be vital when considering new cancer therapies, and would not have been considered using other methods of analysing metabolic profiles. Alternatively, the conserved pathways observed irrespective of HIF-1 function seem to be central in each scenario and therefore targeting these pathways could be potentially the best targets to damage cancer cell metabolism in hypoxia beyond its repair, thus offering sustainable targets for the future.

# References

Baker LCJ, Boult JKR, Walker-Samuel S, Chung YL, Jamin Y, Ashcroft M, Robinson SP (2012) The HIF-pathway inhibitor NSC-134754 induces metabolic changes and anti-tumour activity while maintaining vascular function. Br J Cancer 106(10):1638–1647. doi:10.1038/bjc.2012.131

Brown LM, Cowen RL, Debray C, Eustace A, Erler JT, Sheppard FCD, Parker CA, Stratford IJ, Williams KJ (2006) Reversing hypoxic cell chemoresistance in vitro using genetic and small molecule approaches targeting hypoxia inducible factor-1. Mol Pharmacol 69(2):411–418. doi:10.1124/mol.105.015743

Diaz-Ruiz R, Uribe-Carvajal S, Devin A, Rigoulet M (2009) Tumor cell energy metabolism and its common features with yeast metabolism. Biochim Biophys Acta 1796(2):252–265

Esteban MA, Maxwell PH (2005) HIF, a missing link between metabolism and cancer. Nat Med 11(10):1047–1048

Gordan JD, Thompson CB, Simon MC (2007) HIF and c-Myc: sibling rivals for control of cancer cell metabolism and proliferation. Cancer Cell 12(2):108–113. doi:10.1016/j.ccr.2007.07.006

Griffin JL, Shockcor JP (2004) Metabolic profiles of cancer cells. Nat Rev Cancer 4(7):551–561. doi:10.1038/nrc1390

Griffiths JR, McSheehy PMJ, Robinson SP, Troy H, Chung YL, Leek RD, Williams KJ, Stratford IJ, Harris AL, Stubbs M (2002) Metabolic changes detected by in vivo magnetic resonance studies of HEPA-1 wild-type tumors and tumors deficient in hypoxia-inducible factor-1 beta (HIF-1 beta): evidence of an anabolic role for the HIF-1 pathway. Cancer Res 62(3):688–695

Kanehisa M, Goto S (2000) KEGG: Kyoto Encyclopedia of genes and genomes. Nucleic Acids Res 28(1):27–30. doi:10.1093/nar/28.1.27

Kim J-W, Tchernyshyov I, Semenza GL, Dang CV (2006) HIF-1-mediated expression of pyruvate dehydrogenase kinase: a metabolic switch required for cellular adaptation to hypoxia. Cell Metab 3(3):177–185

Semenza GL (2012) Hypoxia-inducible factors: mediators of cancer progression and targets for cancer therapy. Trends Pharmacol Sci 33(4):207–214. doi:10.1016/j.tips.2012.01.005

Sumner LW, Amberg A, Barrett D, Beale MH, Beger R, Daykin CA, Fan TWM, Fiehn O, Goodacre
    R, Griffin JL, Hankemeier T, Hardy N, Harnly J, Higashi R, Kopka J, Lane AN, Lindon JC,
    Marriott P, Nicholls AW, Reily MD, Thaden JJ, Viant MR (2007) Proposed minimum reporting
    standards for chemical analysis. Metabolomics 3(3):211–221. doi:10.1007/s11306-007-0082-2
Roberts DL, Williams KJ, Cowen RL, Barathova M, Eustace AJ, Brittain-Dissont S, Tilby MJ,
    Pearson DG, Ottley CJ, Stratford IJ, Dive C (2009) Contribution of HIF-1 and drug penetrance
    to oxaliplatin resistance in hypoxic colorectal cancer cells. Br J Cancer 101(8):1290–1297.
    doi:10.1038/sj.bjc.6605311
Ruan K, Song G, Ouyang GL (2009) Role of Hypoxia in the hallmarks of human cancer. J Cell
    Biochem 107(6):1053–1062. doi:10.1002/jcb.22214
Troy H, Chung YL, Mayr M, Ly L, Williams K, Stratford I, Harris A, Griffiths J, Stubbs M
    (2005) Metabolic profiling of hypoxia-inducible factor-1 beta-deficient and wild type Hepa-1
    cells: effects of hypoxia measured by H-1 magnetic resonance spectroscopy. Metabolomics
    1(4):293–303. doi:10.1007/s11306-005-0009-8

# Chapter 6
# Case Study: Systems Biology of Chemotherapy Resistance in Hypoxic Cancer

Current strategies for treatment of cancer include surgery, chemotherapy and radiotherapy (Pelengaris and Khan 2006). Surgery is used to remove the cancerous tumour and is often used in combination with radio- or chemotherapy. Radiotherapy uses ionising radiation directed to kill cancer cells and is often applied to treat tumours localised to one area of the body. Ionising radiation damages DNA, resulting in cell death (Pelengaris and Khan 2006).

Chemotherapy is the treatment of cancer with an anti-neoplastic therapeutic and is often administered as a combination of multiple therapies. Chemotherapy is administered neoadjuvant (pre-operative) to shrink the primary tumour or adjuvant (post-operative) to prevent reoccurrence of the cancer and kill any cancer cells that may have metastasised to other areas of the body (Pelengaris and Khan 2006). Many of these agents act by killing the fast-dividing cancer cells; however they also damage healthy proliferating cells in the bone marrow, digestive tract and hair follicles (Pelengaris and Khan 2006).

## 6.1 Doxorubicin

The development of new cancer drugs has been driven at a rapid pace over the past decades; however a complete understanding of the mechanisms of action for existing, widely used, chemotherapy agents remains to be fully elucidated. Focusing research towards exploring the mechanisms behind anti-tumour activity of frequently used chemotherapies is essential to help identify where these drugs fail and guide the focus of new therapies. In this case study the metabolic response of cancer cells to doxorubicin treatment was investigated to help ascertain why the drug is effective in treating oxygenated cancer cells but fails to be effective in treating solid hypoxic tumours (Sullivan et al. 2006; Sullivan et al. 2008; Cho et al. 2013).

Doxorubicin is a chemotherapeutic and licensed under the trade names Adriamycin® and Rubex®. The compound is an anthracycline antibiotic synthesised from daunorubicin, a naturally occurring product of various wild-type strains of *Streptomyces*. Figure 6.1 shows the chemical structure of doxorubicin, which

E. G. Armitage et al., *Correlation-based network analysis of cancer metabolism,*
SpringerBriefs in Systems Biology, DOI 10.1007/978-1-4939-0615-4_6,
© The Authors 2014

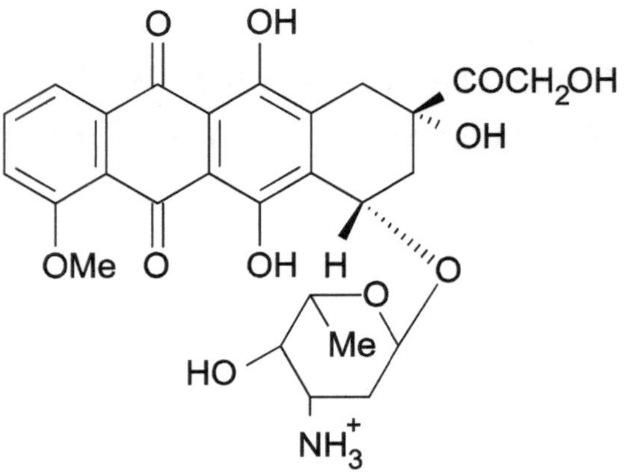

**Fig. 6.1** The chemical
structure of doxorubicin
with molecular formula
$C_{27}H_{29}NO_{11}$ and a molecular
weight 543.52. Doxorubicin
is an anti-neoplastic antibiotic
synthesised from strains of
*Streptomyces*

has the systematic IUPAC name (8S,10S)-10-(4-amino-5-hydroxy-6-methyl-tetrahydro-2H-pyran-2-yloxy)-6,8,11-trihydroxy-8-(2-hydroxyacetyl)-1-methoxy-7,8,9,10-tetrahydrotetracene-5,12-dione.

Passive diffusion transports doxorubicin through the cell membrane (Mayer et al. 1986). A number of mechanisms for the mode of action of doxorubicin have been proposed. Doxorubicin intercalates between the C and G DNA base pairs, causing deformation of the DNA backbone. Binding of doxorubicin to DNA occurs through non-covalent intercalation, where the planar aromatic four-membered ring system of doxorubicin inserts into the major groove of the DNA (Powis and Prough 1987). Doxorubicin is precise in its intercalation point between the terminal CpG, and the ring system inserts left to right, head on, into the major groove (Chaires 1990). The chromophore inserts parallel and between the adjacent base pairs, which does not disrupt the stacking of the base pairs but causes the pairs to spread vertically along the DNA to allow for intercalation. This causes deformation of the DNA backbone and the DNA adopts a long rigid structure. The sugar moiety of doxorubicin bonds to the minor groove of the DNA, which has the greatest anti-tumour action due to the stronger DNA binding properties. Within the minor groove the sugar moiety of doxorubicin bonds to the sugar phosphate backbone and neighbouring proteins (with the later forming the strongest bonds). If the sugar moiety is not present the chromophore would rapidly intercalate and de-intercalate (Fornari et al. 1994). Intercalation disrupts replication and transcription processes.

In aerobic cells doxorubicin produces free radicals resulting in cellular damage which initiates apoptosis. Doxorubicin is catalysed by flavoproteins to produce a semiquinone free radical that can cause cellular damage (Powis and Prough 1987) and further redox cycling of the superoxide ion causes additional damage. Doxorubicin undergoes a one-electron reduction to form a free radical under aerobic conditions (Fig. 6.2).

Additional free radicals formed during oxidative phosphorylation metabolism can interact with doxorubicin and generate the semiquinone compound. Doxorubicin forms the semiquinone intermediate and redox cycling transfers an electron onto oxygen allowing doxorubicin to reform (Scheulen et al. 1982). In low oxygenated

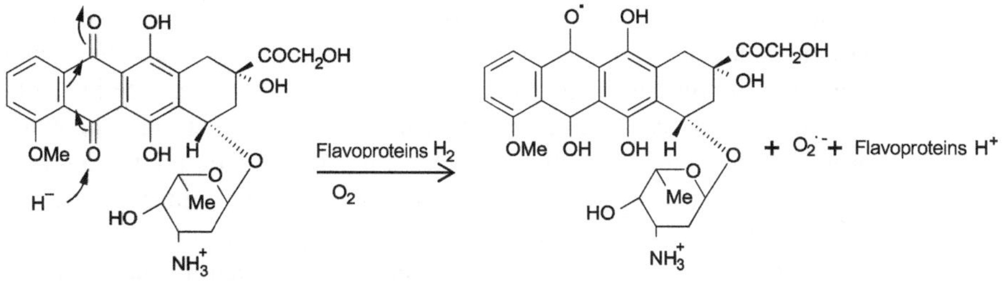

**Fig. 6.2** Reduction of doxorubicin catalysed by flavoproteins to form a semiquinone free radical and superoxide, which causes DNA damage

conditions, this process cycles until oxygen depletes and doxorubicin is rapidly catalysed to 7-deoxyalglycone (Scheulen et al. 1982). The superoxide compound removes a proton from the phenol group to produce a phenolate anion. The intermediate anion is oxidised and generates the semiquinone free radical structure. The semiquinone structure of doxorubicin can form hydroxyl radicals, which are catalysed by Fe(II), to oxidise polyunsaturated fatty acids, degrade deoxyribose and produce DNA double strand breaks (Powis and Prough 1987). Alternatively, doxorubicin and Fe(III) chloride may directly target lipid peroxidation.

Non-cancerous tissue has an alkaline extracellular pH of approximately 7.4; however the pH decreases to around pH 6.5–6.8 in malignant cells (Goode and Chadwick 2008). Increased acidity arises in hypoxic cells as a consequence of non-oxidative phosphorylation metabolism, which subsequently increases the efflux of lactate (Goode and Chadwick 2008). An insufficient oxygen supply causes a shift in metabolism as the oxidation of reduced nicotinamide adenine dinucleotide (NADH) to nicotinamide adenine dinucleotide ($NAD^+$) by the electron transport chain is unable to sustain glycolysis. As a result the $NAD^+$ is restored through reducing pyruvate to lactate. Acidity is generated due to excess protons produced during ATP hydrolysis reducing intracellular pH. During glycolysis a net production of 2 ATP molecules can be used in cellular processes to release protons (Goode and Chadwick 2008). The cells transport protons out from the cancer cell via the ATPase pump or monocarboxylate transporter-1 and 4 to deacidify the cytosol (Le Floch et al. 2011). Lactate and protons are expelled into the extracellular surroundings to maintain a neutral acidity within the intracellular environment. This action can protect the cancer cell from an immune system attack and overcome programmed cell death (Tomiyama et al. 2006). In addition, the acidic extracellular environment damages surrounding cells to facilitate metastasis (Feron 2009). Furthermore, the cell membrane transporter carbonic anhydrase IX (CA-IX) has been shown to be up-regulated in hypoxia by HIF-1 (Chiche et al. 2009) to efflux bicarbonate. Often an overexpression of CA-IX is associated with a poor prognosis as these tumours tend to be highly metastatic and have poor vascularisation (Koukourakis et al. 2001).

The multitude of cellular deacidification mechanisms causes an increase in the extracellular acidity levels. Tumour acidity has been shown to prevent the uptake of doxorubicin into the tumour cells (Swietach et al. 2012). Low extracellular pH reduces the uptake of weakly basic compounds, such as doxorubicin, into the cytosol due to protonation of the drug, which reduces drug accumulation in the tumour. Additionally, weakly basic drugs can be protonated in the cytosol and subsequently diffuse into acidified endosomes reducing the accumulation in the cell nucleus and subsequently reducing cytotoxicity. Preventing extracellular acidity has been reported to increase doxorubicin toxicity (Goode and Chadwick 2008). Greater uptake of doxorubicin has been shown at pH 7.4 compared to pH 6.6 suggesting uptake is pH dependent (Raghunand et al. 2003). Approaches to overcome acidosis in cancer cells have been reported to help improve chemotherapy cytotoxicity. For example, the influx of doxorubicin is increased when the enzyme pyruvate dehydrogenase kinase-3 is inhibited (Lu et al. 2008). Forced expression of pyruvate dehydrogenase kinase-3 caused an increase in lactate accumulation and was correlated with drug resistance. Furthermore, minimising the uptake of doxorubicin into acidic endosomes increase cytotoxicity through combination therapy with omeprazole (Tredan et al. 2007). Omeprazole is a proton pump inhibitor that acts to prevent drug uptake into the acidic endosomes. Initial results showed improved doxorubicin penetration through multilayer cell culture (Tredan et al. 2007).

Many solid tumours express P-glycoprotein (Pgp) to aid their survival when exposed to chemotherapeutics. Expression of Pgp has been shown to be correlated with tumour hypoxia (Comerford et al. 2002). Pgp was observed in Chinese hamster plasma membranes expressed Pgp in ovary cells (Juliano and Ling 1976). Structurally Pgps are inward facing transporters allowing drugs to enter via the cytoplasm and inner leaflet of the lipid bilayer (Aller et al. 2009). Pgps are mainly expressed in areas including the gut, the blood–brain barrier and the blood–testis barrier to facilitate transport of potentially damaging substances (Sharom 2008). The drug efflux pump can significantly reduces the intracellular concentration of xenobiotic compounds such as doxorubicin and hence contribute to drug resistance (Wartenberg et al. 2001). Doxorubicin enters the Pgp chamber on the plasma membrane and the nucleotide-binding domain (NBD) is phosphorylated. There are 12 transmembrane regions available for the drug to bind. To activate the pumping mechanism to export the chemotherapy drug located in the chamber, 2 ATP molecules attach to the ATP-binding cassettes causing dimerisation. Consequently, Pgp undergoes a conformational change at the NBD, initiating structural change in the transmembrane domain to facilitate the release of doxorubicin into the extracellular environment (Aller et al. 2009). ATP hydrolysis releases the drug from the chamber through widening and motioning. Further ATP hydrolysis restores the Pgp to its original configuration releasing inorganic phosphate and ADP. Studies have suggested that the *Mdr-1* gene could be regulated by HIF-1α (Comerford et al. 2002). Additionally, larger tumours have an increased expression of HIF-1α and Pgp, and increasing levels of reactive oxygen species (ROS) have been reported to be associated with a down regulation of the expression of HIF-1α and Pgp in large tumours (Wartenberg et al. 2005). Verapamil and cyclosporin A are inhibitors of the Pgp drug efflux pump (Gottesman et al. 2002) but the compounds developed to date have neurotoxic side effects (Gottesman et al. 2002).

Doxorubicin is a topoisomerase II poison (Fornari et al. 1994). Topoisomerase II is responsible for ligating DNA double strand breaks following DNA replication. A decrease in the expression of topoisomerase II has been reported to be associated with a reduction in the sensitivity to chemotherapeutics (Ogiso et al. 2000; Kang et al. 1996). The proteasome inhibitor lactacystin overcomes hypoxia resistance due to topoisomerase II through preventing topoisomerase II depletion (Ogiso et al. 2000). This enables the chemotherapeutics to have a greater cytotoxic effect to hypoxic tumours.

## 6.2   Network-Based Correlation Analysis to Study Hypoxia Induced Chemoresistance

Systems biology is the study of the complex interactions within a biological system to determine the emergent properties arising from the interaction of the genes, proteins and metabolites. This can be considered as a holistic approach and presents an understanding into how interactions give rise to a system's function and behaviour. Cancer is a complex disease that arises due to a series of mutations, not a single gene effect, and it is the combination of these gene mutations that give rise to cancer as an emergent property of the cellular system that it interacts with. Hanahan and Weinburg proposed six biological hallmarks of cancer that arise as a result of genome instability highlighting the complexity of the disease (Hanahan and Weinberg 2000). The proposed hallmarks of cancer include sustaining proliferative signalling, avoiding growth suppressors, opposing cell death, replicating indefinitely, angiogenesis and the ability to metastasise into previously non-cancerous regions. Two additional hallmarks of cancer have recently been recognised, which include evading immune destruction and, importantly, reprogramming of energy metabolism (Hanahan and Weinberg 2011). Applying systems biology to cancer research offers a new perspective and has the potential to unveil the emergent properties of cancer metabolism.

The aim of this case study was to identify strongly correlated metabolites within MDA-MB-231 breast cancer cells exposed to a low-oxygen environment and during treatment with doxorubicin. Correlation differences were interpreted by isolating the interconnecting metabolites in the human metabolic network in order to construct a metabolic network of chemoresistance. Subsequently, novel chemotherapeutic strategies were determined from the constructed networks. In order to reveal the underlying systemic biological response, patterns of correlations were observed within an obtained gas chromatography-mass spectrometry (GC-MS) dataset from the analysis of cells exposed to a range of oxygen potentials (as described in the previous chapter as normoxia, hypoxia and anoxia) and treated with different concentrations of doxorubicin. It is thought that even small variations in enzymatic reactions can cause significantly different correlations when comparing sample treatments. The differences in metabolite concentrations can be described by pairwise correlations that are indicative of the mechanism of hypoxia-induced chemotherapy resistance. Determining the shortest path connecting pairwise correlations will produce a network hoped to describe the global property of the systems response.

## 6.2.1   Correlation Analysis

Three metabolic networks were created to describe the cellular response of MDA-MB-231 cancer cells: (a) treated with a cytotoxic dose of doxorubicin in high oxygen levels, (b) treated with a non-cytotoxic dose of doxorubicin in low oxygen levels, and (c) treated with a cytotoxic dose of doxorubicin in low oxygen levels. These networks offer an insight into how cells respond, metabolically, to environmental perturbations and further reveal metabolic pathways as potential targets to help overcome hypoxia-induced chemotherapy resistance. Through observing the networks of resistance and overcoming resistance by treating with a higher drug dose, the identification of novel therapeutic targets to improve cytotoxicity in resistant cells was possible.

Using a systems biology approach, the metabolic network of chemotherapy resistance was identified using network-based correlation analysis. Construction of metabolic networks, through compiling all relevant pathway responses, enables a greater insight into the network mechanisms. Thus, the topology and dynamics (fluctuations) of the metabolic responses were explored. Furthermore, these networks were cross compared to investigate the pathway responses that overcome chemoresistance.

Using MDA-MB-231 cells exposed to three levels of oxygen normoxia, hypoxia and anoxia (21, 1 and 0 % $O_2$ respectively) and three levels of drug treatment (0, 0.1 and 1 μM), with 30 repeat measurements, a total of 52 metabolites were identified using GC-MS analysis. Comparisons between therapeutic doses of doxorubicin (0.1 μM) administered at 21 % oxygen levels yielded a Pearson's correlation that significantly differed with drug treatment at a level of $α = 0.05$. The pairwise correlation coefficient between fructose and glutamate for untreated cells was $-0.27$ whereas for treated cells was 0.73. The reversal in the metabolic correlation may suggest a response in the regulation of the underlying metabolic pathway.

Pairwise correlations were subsequently identified for normoxia samples treated with and without 1 μM doxorubicin. Table 6.1a contains examples of the most significant differential correlations. This dose of doxorubicin was shown to be highly toxic to the cells in normoxic conditions and therefore these correlations represent high toxicity in the cells. These mechanisms do not reflect a clinical dose response to doxorubicin; however, they do provide an insight into the cellular response to toxicity. Clinically, a dose accumulation of 550 mg/m² results in unwanted cardiac

**Table 6.1** The most significantly different pairwise correlations of metabolites for MDA-MB-231 cells comparing cells dosed with doxorubicin to cells not dosed. **a** Normoxia with or without 1 μM dox, **b** hypoxia with or without 0.1 μM dox, and **c** anoxia with or without 0.1 μM dox

| Metabolite A | Metabolite B | Difference | $r$ (N) | $r$ (N+dox) |
|---|---|---|---|---|
| *Normoxia (N) with or without 1 μM doxorubicin (dox)* | | | | |
| Malate | Lactate | 1.04 | −0.20 | 0.84 |
| Malate | Pyruvate | 0.96 | −0.14 | 0.82 |
| Malate | Threitol/erythritol | 0.74 | 0.14 | 0.88 |
| Xylitol | Malate | 0.69 | 0.21 | 0.90 |
| Benzoic acid | Isoleucine | 0.57 | 0.73 | −0.16 |
| Hypotaurine | Glycerol | 0.54 | 0.23 | 0.77 |
| Glutamate | Malate | 0.49 | 0.40 | 0.89 |
| Sorbose/fructose | Malate | 0.47 | 0.23 | 0.70 |
| Scyllo-inositol/myo-inositol/ inositol | Malate | 0.42 | 0.40 | 0.82 |
| Allose/mannose/galactose/ glucose | Malate | 0.42 | 0.72 | 0.30 |
| *Metabolite A* | *Metabolite B* | *Difference* | *r (H)* | *r (H + dox)* |
| *Hypoxia (H) with or without 0.1 μM doxorubicin (dox)* | | | | |
| Benzoic acid | Lactate | 0.47 | 0.27 | 0.74 |
| Sorbose/fructose | Benzoic acid | 0.46 | 0.25 | 0.71 |
| Glutamate | Lactate | 0.44 | 0.40 | 0.84 |
| *Metabolite A* | *Metabolite B* | *Difference* | *r (A)* | *r (A + dox)* |
| *Anoxia (A) with or without 0.1 μM doxorubicin (dox)* | | | | |
| Octadecanoic acid | Glutamate | 0.8 | −0.03 | 0.77 |
| Octadecanoic acid | Isoleucine | 0.79 | 0.01 | 0.80 |
| Octadecanoic acid | Threonine | 0.75 | −0.01 | 0.74 |
| Octadecanoic acid | Threitol/erythritol | 0.65 | 0.16 | 0.81 |
| Glutamate | 4-hydroxyproline | 0.62 | 0.16 | 0.78 |
| Octadecanoic acid | Sorbitol/galactose/ glucose | 0.61 | 0.14 | 0.75 |
| Octadecanoic acid | Glycerol | 0.57 | 0.18 | 0.75 |
| Octadecanoic acid | Sorbose/fructose | 0.55 | 0.21 | 0.76 |
| 4-hydroxyproline | Isoleucine | 0.48 | 0.28 | 0.76 |
| Octadecanoic acid | Aspartate | 0.48 | 0.25 | 0.73 |

damage, which can result in death and is characterised by a decrease in mitochondrial oxidative phosphorylation and reactive oxygen species (ROS) damage myocytes (Ferrari et al. 1993).

Pearson's correlation for cells exposed to hypoxia and treated with and without 0.1 μM doxorubicin yielded three significantly different pairwise correlations (Table 6.1b). Furthermore, Pearson's correlation analysis between anoxia and anoxia samples treated with 0.1 μM doxorubicin yielded 10 metabolic correlations (Table 6.1c). The resistant response to a low drug dose of doxorubicin appears to have a greater effect on the metabolism than that observed for drug treated cells cultured in normal oxygen levels. This may be due to low-oxygen induced resistance being regulated through metabolism.

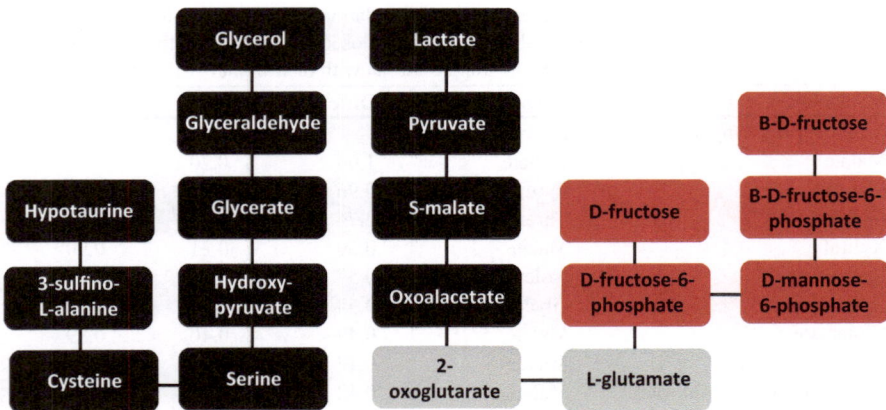

**Fig. 6.3** Network of the shortest path connecting pairwise correlations of metabolites that significantly differed for MDA-MB-231 cells cultured in normoxia compared to normoxia dosed with 0.1 μM or 1 μM doxorubicin. Shortest pathways connecting two metabolites were determined using Edinburgh human metabolic network (EHMN). The *red* nodes represent pathways of metabolites present in the lower drug dose, *black* represents pathways of metabolites present in higher drug dose and *grey* are the pathways of metabolites present in both doses of doxorubicin

## 6.2.2   Network Analysis

Pairwise metabolites that differed significantly in response to drug treatment were isolated within the EHMN and the shortest pathway connecting two metabolites was acquired. Pathway responses of all drug doses were combined to generate a single network of drug response in normoxic cells (Fig. 6.3). This is a network of all the potential mechanisms of drug action with respect to metabolism for cells exposed to a normal oxygen tension. Shared components between both drug doses are displayed as grey nodes. Additionally, low drug dose pathways are shown as red nodes and high drug dose pathways are shown in black.

Figure 6.3 describes glutamine and 2-oxoglutarate as conserved features of the metabolic response to doxorubicin irrespective of the dose. These metabolites connect in a single pathway; however the metabolic regulation of the metabolites may be the result of a dose dependent pathways response. Glutamine and 2-oxoglutarate have a role in the TCA cycle, which is part of central carbon metabolism. Doxorubicin may therefore be targeting energy metabolism through dose dependent mechanisms. For example, the lower drug dosed response is through a pathway connecting to fructose metabolism. In contrast, the higher dose response occurs through a pathway connecting to lactate. Furthermore, an additional pathway response from the higher drug dose occurs through the pathway connecting glycerol and hypotaurine. A combination of these two pathway responses may be the result of pathway the higher cytotoxicity. Hypotaurine and metabolic precursors in this additional pathway, such as cysteine, are antioxidants. A high dose of doxorubicin is expected to generate ROS when doxorubicin is catalysed by flavoproteins to produce a semi-quinone free radical (Powis and Prough 1987). Furthermore, redox cycling of the

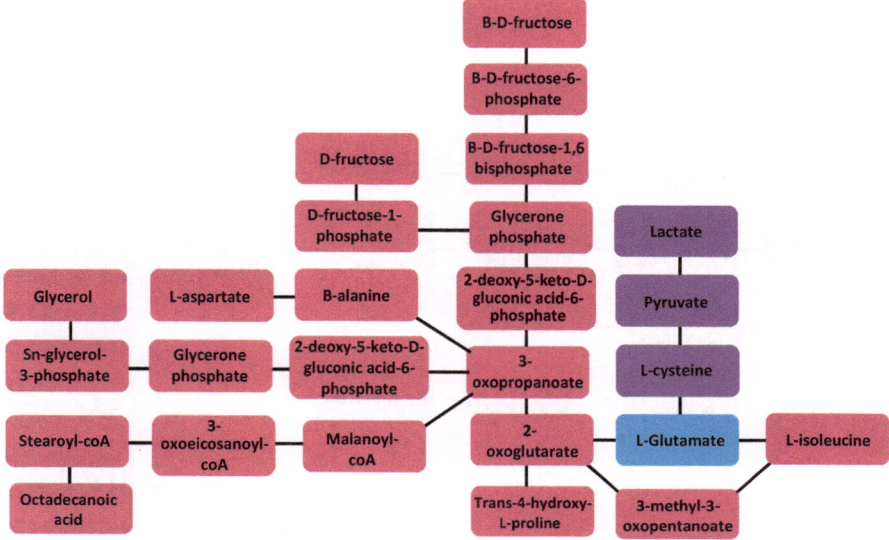

**Fig. 6.4** Network of the shortest path connecting pairwise correlations of definitively identified metabolites that significantly differed for MDA-MB-231 cells cultured in hypoxia treated with or without 0.1 μM and MDA-MB-231 cells cultured in anoxia treated with or without 0.1 μM doxorubicin. The shortest pathways connecting two metabolites were determined using Edinburgh human metabolic network (EHMN). The *purple* nodes represent pathways of metabolites present in the hypoxia drug treated cells, *pink* represents pathways of metabolites present in anoxia drug treated cells and *blue* are the pathways of metabolites present in both oxygen conditions dosed with drug

superoxide ion causes additional ROS generation. Metabolic regulation may cause a response in hypotaurine and cysteine to minimise the cytotoxic effects of drug-induced generation of ROS at higher doses (Aruoma et al. 1988).

Network-based correlation analysis was applied to construct the network shown in Fig. 6.4. Pathways relating to hypoxia drug response are represented as purple nodes and pathways relating to anoxia drug response are represented as pink nodes. Conserved features drug response to both oxygen tensions in the network are displayed as blue nodes. The unique node conserved between the dose responses to the two oxygen tensions was L-glutamate.

L-glutamate interacts with the metabolic network through hypoxic and anoxic specific pathways, such as through L-cysteine, pyruvate and lactate for hypoxia and 2-oxoglutarate and trans-4-hydroxyl-L-proline for anoxia. The hypoxia-related pathway is a central carbon metabolism pathway suggesting hypoxic cells are changing fluxes through energy metabolism to mediate resistance. In comparison, anoxic pathways are directing the metabolic response though a series of metabolites to have a large change in the pathways directed towards octadecanoic acid (as described by edge thickness in the network). One of the metabolites in this pathway is malanoyl-CoA, which is known to be a building block for fatty acid synthesis and is a regulator of mitochondrial fatty acid synthesis. Fatty acid synthesis has been

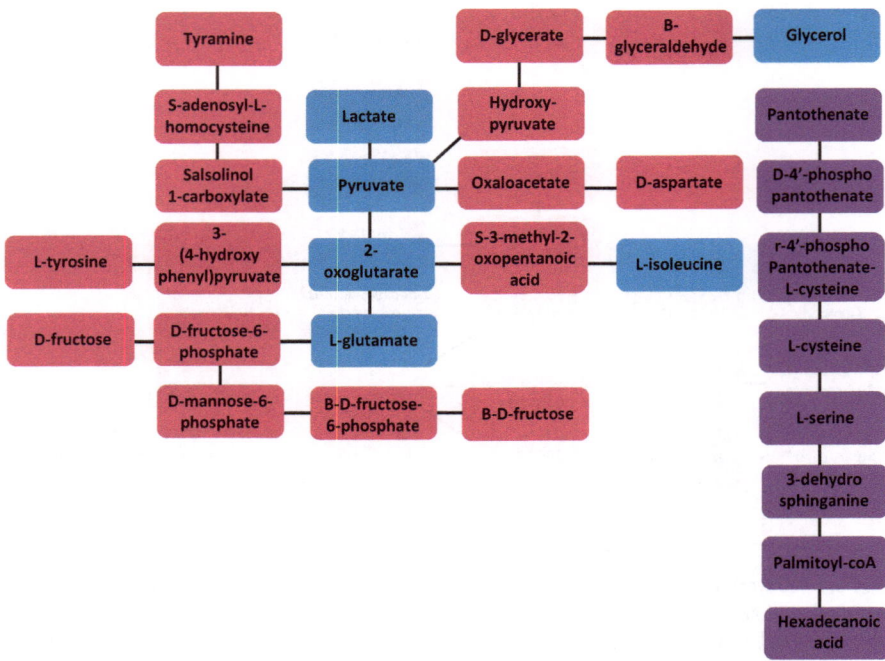

**Fig. 6.5** Network of the shortest path connecting pairwise correlations of definitively identified metabolites that significantly differed for MDA-MB-231 cells cultured in hypoxia compared to hypoxia treated with 1 μM and MDA-MB-231 cells cultured in anoxia to anoxia treated with 1 μM doxorubicin. Shortest pathways connecting two metabolites were determined using Edinburgh human metabolic network (EHMN). The *purple* nodes represent pathways of metabolites present in the hypoxia drug treated cells, *pink* represents pathways of metabolites present in anoxia drug treated cells and *blue* are the pathways of metabolites present in both oxygen conditions dosed with drug

implicated with numerous human tumours and inhibiting the fatty acid synthase (FASN) enzyme has been suggested to suppress tumour growth (Kuhajda 2000; Swinnen et al. 2000). The pathway connecting malanoyl-CoA to octadecanoic acid may therefore be a potential therapeutic target to help overcome low-oxygen induced chemoresistance.

Figure 6.5 shows the metabolic response to cytotoxic doses of doxorubicin in low oxygenated samples and it highlights that metabolites L-glutamate, 2-oxoglutarate, pyruvate and lactate are conserved and are interconnected in a single pathway for hypoxia; however they are interconnected with other metabolites for anoxia cells. The anoxic response connects metabolites described above to glycerol, aspartate, fructose and tyramine/tyrosine which are metabolites involved in central carbon metabolism. Lactate and pyruvate, which are also metabolites of central carbon metabolism, were also mapped in seven of the pathway reconstructions. A cytotoxic dose of doxorubicin may be increasing the enzymatic activity of LDH, the flux of which may be supported by metabolites such as glycerol. This suggests the pyruvate and lactate pathway is an effective therapeutic target to overcome hypoxia-induced resistance to chemotherapy.

## 6.3 Conclusion

Systems biology is the study of the complex interactions within a biological system to determine the emergent properties arising from the interaction of the genes, proteins and metabolites. Applying systems biology to cancer research offers a new perspective and has the potential to unveil the emergent properties of cancer metabolism.

As highlighted in the previous chapter, pathways are often compounded of different features within "traditional" pathways and have been shown in both case studies to provide an alternative way of viewing cancer metabolism. The sub-networks showing the change in pathway regulation caused by lowering the oxygen microenvironment cells are exposed to enables a clearer understanding of cancer metabolism in low oxygen environments in general, and, more specifically, with respect to HIF-1 function and chemoresistance to doxorubicin. With knowledge both of the cancer cell's genetic influence in low oxygen survival and the reasons for which current chemotherapies are not so successful in this environment, it has been possible to reveal alternative pathways that could be useful in targeting for future therapies.

## References

Aller SG, Yu J, Ward A, Weng Y, Chittaboina S, Zhuo R, Harrell PM, Trinh YT, Zhang Q, Urbatsch IL, Chang G (2009) Structure of P-glycoprotein reveals a molecular basis for poly-specific drug binding. Science 323(5922):1718–1722. doi:10.1126/science.1168750

Aruoma OI, Halliwell B, Hoey BM, Butler J (1988) The antioxidant action of taurine, hypotaurine and their metabolic precursors. Biochem J 256(1):251–255

Chaires JB, Herrera JE, Woring M (1990) Biochemistry 29:6145–6153

Chiche J, Ilc K, Laferriere J, Trottier E, Dayan F, Mazure NM, Brahimi-Horn MC, Pouyssegur J (2009) Hypoxia-inducible carbonic anhydrase IX and XII promote tumor cell growth by counteracting acidosis through the regulation of the intracellular pH. Cancer Res 69(1):358–368. doi:10.1158/0008-5472.can-08-2470

Cho K, Shin H-W, Kim Y-I, Cho C-H, Chun Y-S, Kim T-Y, Park J-W (2013) Mad1 mediates hypoxia-induced doxorubicin resistance in colon cancer cells by inhibiting mitochondrial function. Free Radic Biol Med doi:10.1016/j.freeradbiomed.2013.02.022

Comerford KM, Wallace TJ, Karhausen J, Louis NA, Montalto MC, Colgan SP (2002) Hypoxia-inducible factor-1-dependent regulation of the multidrug resistance (MDR1) gene. Cancer Res 62(12):3387–3394

Feron O (2009) Pyruvate into lactate and back: from the Warburg effect to symbiotic energy fuel exchange in cancer cells. Radiother Oncol 92(3):329–333. doi:10.1016/j.radonc.2009.06.025

Ferrari E, Taillan B, Dujardin P, Morand P (1993) Cardiotoxicity of anthracyclines—clinical-features, incidence, monitoring. Presse Med 22(21):999–1004

Fornari FA, Randolph JK, Yalowich JC, Ritke MK, Gewirtz DA (1994) Interference by doxorubicin with DNA unwinding in MCF-7 breast-tumour cells. Mol Pharmacol 45(4):649–656

Goode JA, Chadwick DJ (2008) pH, hypoxia and metastasis. The tumour microenvironment: causes and consequences of hypoxia and acidity. Wiley, New York. doi:10.1002/0470868716.ch11

Gottesman MM, Fojo T, Bates SE (2002) Multidrug resistance in cancer: role of ATP-dependent transporters. Nat Rev Cancer 2(1):48–58. doi:10.1038/nrc706

Hanahan D, Weinberg RA (2000) The hallmarks of cancer. Cell 100 (1):57–70. doi:10.1016/s0092-8674(00)81683-9

Hanahan D, Weinberg RA (2011) Hallmarks of cancer: the next generation. Cell 144(5):646–674. doi:10.1016/j.cell.2011.02.013

Juliano RL, Ling V (1976) Surface glycoprotein modulating drug permeability in Chinese-hamster ovary cell mutants. Biochimica Biophysica Acta 455 (1):152–162. doi:10.1016/0005-2736(76)90160-7

Kang Y, Greaves B, Perry RR (1996) Effect of acute and chronic intermittent hypoxia on DNA topoisomerase II alpha expression and mitomycin C-induced DNA damage and cytotoxicity in human colon cancer cells. Biochem Pharmacol 52 (4):669–676. doi:10.1016/0006-2952(96)00343-7

Koukourakis MI, Giatromanolaki A, Sivridis E, Simopoulos K, Pastorek J, Wykoff CC, Gatter KC, Harris AL (2001) Hypoxia-regulated carbonic anhydrase-9 (CA9) relates to poor vascularization and resistance of squamous cell head and neck cancer to chemoradiotherapy. Clin Cancer Res 7(11):3399–3403

Kuhajda FP (2000) Fatty-acid synthase and human cancer: new perspectives on its role in tumor biology. Nutrition 16 (3):202–208. doi:10.1016/s0899-9007(99)00266-x

Le Floch R, Chiche J, Marchiq I, Naiken T, Ilk K, Murray CM, Critchlow SE, Roux D, Simon M-P, Pouyssegur J (2011) CD147 subunit of lactate/H+symporters MCT1 and hypoxia-inducible MCT4 is critical for energetics and growth of glycolytic tumors. Proc Natl Acad Sci U S A 108(40):16663–16668. doi:10.1073/pnas.1106123108

Lu C-W, Lin S-C, Chen K-F, Lai Y-Y, Tsai S-J (2008) Induction of pyruvate dehydrogenase kinase-3 by hypoxia-inducible factor-1 promotes metabolic switch and drug resistance. J Biol Chem 283(42):28106–28114. doi:10.1074/jbc.M803508200

Mayer LD, Bally MB, Cullis PR (1986) Uptake of adriamycin into large unilamellar vesicles in response to a pH gradient. Biochimt Biophys Acta 857 (1):123–126. doi:10.1016/0005-2736(86)90105-7

Ogiso Y, Tomida A, Lei SH, Omura S, Tsuruo T (2000) Proteasome inhibition circumvents solid tumor resistance to topoisomerase II-directed drugs. Cancer Res 60(9):2429–2434

Pelengaris S, Khan M (eds) (2006) The molecular biology of cancer. Blackwell Publishing Ltd, Oxford

Powis G, Prough RA (eds) (1987) Metabolism and action of anti-cancer drugs. Taylor and Francis, London

Raghunand N, Mahoney BP, Gillies RJ (2003) Tumor acidity, ion trapping and chemotherapeutics I. pH-dependent partition coefficients predict importance of ion trapping on pharmacokinetics of weakly basic chemotherapeutic agents. Biochem Pharmacol 66 (7):1219–1229. doi:10.1016/s0006-2952(03)00468-4

Scheulen ME, Kappus H, Nienhaus A, Schmidt CG (1982) Covalent protein-binding of reactive adriamycin metabolites in rat-liver and rat-heart microsomes. J Cancer Res Clin Oncol 103(1):39–48. doi:10.1007/bf00410304

Sharom FJ (2008) ABC multidrug transporters: structure, function and role in chemoresistance. Pharmacogenomics 9(1):105-127. doi:10.2217/14622416,9.1.105

Sullivan R, Frederiksen LJ, Pare GC, Graham CH (2006) Hypoxia-Inducible Factor 1 is required for hypoxia-induced resistance to doxorubicin in breast carcinoma cells. In: AACR meeting abstracts, 2006, 1, p 403

Sullivan R, Paré GC, Frederiksen LJ, Semenza GL, Graham CH (2008) Hypoxia-induced resistance to anticancer drugs is associated with decreased senescence and requires hypoxia-inducible factor-1 activity. Mol Cancer Ther 7(7):1961–1973

Swietach P, Hulikova A, Patiar S, Vaughan-Jones RD, Harris AL (2012) Importance of intracellular pH in determining the uptake and efficacy of the weakly basic chemotherapeutic drug, doxorubicin. Plos One 7(4):e35949. doi:10.1371/journal.pone.0035949

Swinnen JV, Vanderhoydonc F, Elgamal AA, Eelen M, Vercaeren I, Joniau S, Van Poppel H, Baert L, Goossens K, Heyns W, Verhoeven G (2000) Selective activation of the fatty acid synthesis pathway in human prostate cancer. Int J Cancer 88(2):176-179. doi:10.1002/1097-0215(20001015)88:2<176::aid-ijc5>3.0.co;2-3

Tomiyama A, Serizawa S, Tachibana K, Sakurada K, Samejima H, Kuchino Y, Kitanaka C (2006) Critical role for mitochondrial oxidative phosphorylation in the activation of tumor suppressors Bax and Bak. J Natl Cancer Inst 98(20):1462–1473. doi:10.1093/jnci/djj395

Tredan O, Galmarini CM, Patel K, Tannock IF (2007) Drug resistance and the solid tumor microenvironment. J Natl Cancer Inst 99(19):1441–1454. doi:10.1093/jnci/djm135

Wartenberg M, Ling FC, Schallenberg M, Baumer AT, Petrat K, Hescheler J, Sauer H (2001) Downregulation of intrinsic P-glycoprotein expression in multicellular prostate tumor spheroids by reactive oxygen species. J Biol Chem 276(20):17420–17428. doi:10.1074/jbc.M100141200

Wartenberg M, Gronczynska S, Bekhite MM, Saric T, Niedermeier W, Hescheler J, Sauer H (2005) Regulation of the multidrug resistance transporter P-glycoprotein in multicellular prostate tumor spheroids by hyperthermia and reactive oxygen species. Int J Cancer 113(2):229–240. doi:10.1002/ijc.20596